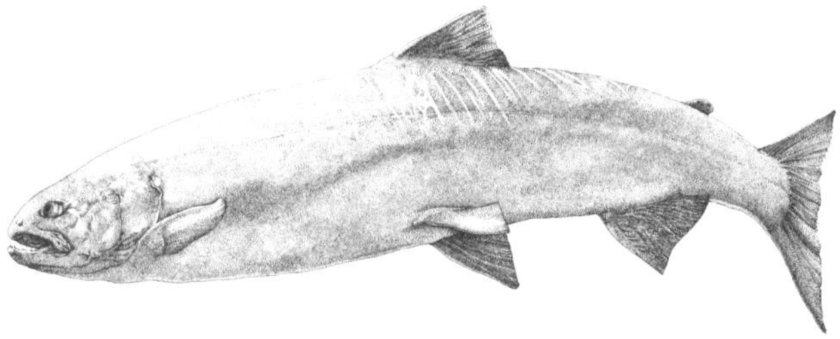

Books by Roderick Haig-Brown

1931 / Silver
1932 / Pool and Rapid
1934 / Panther
1939 / The Western Angler
1941 / Return to the River
1942 / Timber
1943 / Starbuck Valley Winter
1946 / A River Never Sleeps
1948 / Saltwater Summer
1949 / On the Highest Hill
1950 / Measure of the Year
1951 / Fisherman's Spring
1954 / Fisherman's Winter
1954 / Mounted Police Patrol
1956 / Captain of the Discovery
1959 / Fisherman's Summer
1960 / The Farthest Shore
1961 / The Living Land
1962 / Fur and Gold
1962 / The Whale People
1964 / Primer of Fly-Fishing
1964 / Fisherman's Fall
1974 / The Salmon

Bright Waters, Bright Fish

Roderick Haig-Brown

BRIGHT WATERS, BRIGHT FISH

An Examination of Angling in Canada

DOUGLAS & McINTYRE
Vancouver

In association with the British Columbia Wildlife Federation

Published in Canada by
DOUGLAS & McINTYRE LTD.
1615 Venables Street
Vancouver, British Columbia
in association with the British Columbia Wildlife Federation
ISBN 0-88894-284-2

Published in the United States by
TIMBER PRESS
P.O. Box 92
Forest Grove, Oregon
ISBN 0-917304-59-4

Published in co-operation with the Department of Fisheries and Oceans Canada and the Canadian Government Publishing Centre, Supply and Services Canada. Catalogue no. FS99-28-1980E

Canadian Cataloguing in Publication Data

Haig-Brown, Roderick L., 1908-1976.
Bright waters, bright fish

Bibliography: p.
ISBN 0-88894-284-2

1. Fishing - Canada. 2. Fishery management - Canada. I. Title.
SH571.H34 799.1'0971 C80-091174-1

Design by Nancy Legue
Typesetting by Vancouver Typesetting Company Ltd.
Colour separations, printing and binding by
Evergreen Press Ltd.
Printed and bound in Canada.

Contents

Acknowledgements

The editors wish to thank the many people across Canada who gave so generously of their knowledge and time during the development of this book, especially Valerie Haig-Brown and Archie Tuomi of the Department of Fisheries and Oceans, Canada.

Grateful acknowledgement is also made to the following individuals and government departments for their assistance in providing illustrations: Robert Fish, whose fine drawings and watercolours were commissioned by the Department of Fisheries and Oceans, Canada; the Canadian Government Office of Tourism, who provided the majority of the photographs; the *Atlantic Salmon Journal;* Den Austin; Department of Fisheries and Oceans, Canada; George Gruenefeld; Richard Olmstead; Christopher Springmann, and Travel Manitoba, all of whom provided additional photographs. Full credit is given on page 143.

Introduction

A river is a tangible and dramatic force, whether it winds through the spruce and hardwood covered ridges of the Laurentian Shield, the tundra of the north or the mountainous rain forests of the west coast. It is part of a vast mysterious entity that reaches from the gravel of the river bottom and the earth along its banks to the tops of the highest trees, and beyond to the wind currents thousands of feet above which carry insects to destinations unknown to man.

This pervading spirit is the living force of the organism we call earth. And among the most fascinating forms of life in boundless nature are the fishes. They live in a world mostly hidden from man's prying eyes, eagerly sought by anglers who must guess at their habits and whereabouts.

Roderick Haig-Brown was an angler, but one who was not satisfied just to stand on the banks of a river and speculate — he donned a wet suit and mechanical gills to join the fish in their element. And much more than an accomplished angler artful in fooling fish with feathers, he was a first-hand observer, a philosopher who gave meaning to his observations, and a consummate writer who provoked and entertained millions with his articles and books.

Innately modest, Haig-Brown tried to decline inheriting the mantle of Izaak Walton, but to no avail. *Life* magazine described him as "the most eloquent of modern-day fishing prose writers" — as the readers of his more than twenty books already knew. When he remarked of Winchester Cathedral, "within lie the bones of our father, Izaak Walton," it only confirmed his kinship with his illustrious predecessor which his readers had already recognized. But beyond attributes which by themselves would have distinguished other men, Roderick Haig-Brown combined a fierce independence with a soul-searching commitment to duty and public service. In matters of conscience concerning the relationship of man with the natural world he loved and identified with, he spared no one, least of all himself.

The principles Haig-Brown delineates in this book apply not only to fishes and fishing but to all forms of life touched by man. What he says here is vital to our thinking about the preservation and development of all our resources. Man has no divine right to do what he likes with nature, nor can he improve on it. At best, he can bend it and shape it to his needs, and if he does this with care and consideration, his rewards will reach as far as time and compound themselves in the process.

In his writing Roderick Haig-Brown demonstrated the validity of this philosophy over and over again. His literary peers recognized him as an artist willing and able to illuminate his subjects and to extend the average man's understanding. Although he once spoke of fishermen as being loafers and purveyors of tall tales, he also said that they were searchers for experiences and were thoughtful, vigorous, often poetic people.

He once said, "It is fashionable to call any occupation that does not contribute in some dull way to the world's material wealth an 'escape.' It is a ridiculous fashion, as little connected with reality as acute insanity. Reality, for any properly constituted society, can never consist solely in materially productive work. Sports like hunting and fishing, actively and positively followed, are an important and integral part of living for many men. . . ."

A sympathetic left-wing friend once wondered what would become of men like Roderick Haig-Brown in a socialist revolution and questioned whether or not a writer was a producer of some sort. Imagining a liquidation squad for hapless authors, Haig-Brown expostulated, "Producer hell! A writer is a natural resource!"

In his own inimitable way, his life proved it.

When he first saw the Campbell River on Vancouver Island, it was a wild river. Between the unclimbable barrier of Elk Falls and the ocean, a distance of only a few miles, was a prime spawning ground for steelhead, coho and spring salmon. On a bench above high-water mark overlooking the river and not far from its mouth, the naturalist built his home. Here he and his wife Ann raised their family and he wrote his books, stories and articles.

The location afforded him plenty of opportunity to pursue his passion for angling and for observing the fish and other life linked with the stream, from tiny nymphs clinging to rocks on its bottom to the eagles swinging in lazy circles high against the clouds. It was here that the spirit of this remarkable man joined with that of the river and he was granted an understanding that transcended most people's experience. It shone in all his books. It put fire in his lectures. It was a never-ending drive and inspiration in his fight to preserve British Columbia's wild rivers from the destroying dams of the hydroelectric engineers. The fight was lifelong and fraught with disappointments: many battles seemingly won had to be fought again and again.

Because of the quiet strength of his personality and convictions, and the fine-honed edge of his words, people listened carefully to him even when they disagreed. Some might have cursed his interference in their grand schemes, but they respected him. To many he became a living legend.

Years ago I visited him in his home and came away knowing that I had been privileged to meet a great man. There was such a power and dignity and wisdom in him, and he proffered such a warm-hearted friendship, that the evening was unforgettable.

He once wrote, "If one has to die, I should think November would be the best month for it. I should think there is nothing very bad about dying except for the people one has to leave and the things one hasn't had the time to do. When the time comes, if I know what it is all about, I suppose I shall think, among other things, of the fish I haven't caught and the places I haven't fished." He picked November, the beginning of slack time for the angler.

Roderick Haig-Brown died in October 1976 only a few weeks after he finished this book. At his funeral a woman came up to his grieving family and through her tears said, "Whoever will look after the Fraser River now?"

Who will look after all the waters that in Canada still run wild and clear? Roderick Haig-Brown, by the example of his fight to keep our waters clean and a living part of our heritage, and through this book, has charged the people of Canada and especially fishermen with that duty. It is up to us, with courage and conviction, to ensure that the silvery hordes following their destiny in our rivers continue to do so. If we fail, we will have lost much of the beauty and excitement in our part of the natural world. We must carry Haig-Brown's challenge to a successful conclusion. We owe it to the country and to ourselves.

And to good fishing!

Andy Russell
Waterton Lakes National Park, Alberta
July 1980

Nimpkish River, British Columbia

Chapter One The Nature of the Question

Nearly forty years ago I concluded *The Western Angler,* a book about sport fishing in the west with a chapter called "The People's Right to Good Fishing." Throughout the book I had emphasized the duties and responsibilities that anglers owe their resource — the fish and their habitat. In this final chapter I felt it was time to point out the state's responsibility to the angler and to the resource because it was clear, even then, that steadily increasing leisure was a fact of western society and that angling was one of the most popular forms of recreation on the North American continent.

People everywhere seem to enjoy going fishing when they have the opportunity. The sport has long been popular in Russia and China and so far as I know it still is. There is fishing almost everywhere in Europe, from the River Arno in Italy to the salmon rivers of Lapland and Iceland, from Czechoslovakia and Yugoslavia in the east to the trout streams of Brittany in the west. There is fishing in Britain and Spain, in the rivers of Egypt and the rivers of India. There are, I was told by Jean-Louis Gaudet of the Food and Agriculture Organization, more than fifteen million anglers in Europe. For the most part it is people like these who settled North America and they found a continent of rivers and lakes open to fishing. As soon as they had settled down and had time to look around they began to fish.

Interestingly, the kinds of fish and their general behaviour in angling terms are such that European experience has fairly broad application throughout Canada. This means that Canadians inherit, in significant degree, the great wealth of European angling literature and can adopt and build on many of the traditions that have developed to give high meaning to the sport. At the same time, Canadian experience can be enriched by interchange with the parallel developments in the sport among fishermen of the United States, who share the same literature and traditions. All sport depends in the end on traditions, ethics and standards that enrich, without inhibiting, personal experience. The quality of personal experience depends upon subjective values and a host of intangibles, and no sport makes more of these than does angling, no sport offers more scope for highly individualized experience. But shared human experiences — especially in the permanent literature — shared ethics, shared traditions must always give added meaning and value to the keenest experience, those most delightful or most exciting of unexpected happenings that reward the angler so generously.

When I began to fish in North America, in the 1920s and 1930s, it seemed that most Americans and Canadians one met, in the west at all events, had been fishing at one time or another or else planned to try it soon. More than a few were dyed-in-the-wool enthusiasts and among these was a fair proportion of creative experts. In the face of all this interest there was very little regulation or control of the various fisheries, little real knowledge of the fish or their needs, and very little research going on. Most of the available solid information about game fish and their habits was a by-product of commercial fisheries research.

Forty and fifty years ago Canada was a young country and natural resources were not much considered, except as a means of making money. The sport fishery was looked upon as minor and somewhat frivolous, something of an asset to the tourist industry, perhaps, but not much more. Most sport fishing takes place in fresh water, and the various interpretations of the British North America Act, Canada's constitution, make it clear that while the federal government has the general power to regulate, protect and preserve all fisheries, "the exclusive right of fishing exists in the Crown in right of the province": that is, the authority to issue licences or grant leases to fish in nontidal waters flowing through Crown lands belongs to the provinces. This divided jurisdiction produced both confusions and lapses until after World War II, when the nation began to appreciate the importance of its sport fisheries rather more fully. In general, the federal authorities have been inclined to leave the management of freshwater fisheries pretty much to the provinces, simply setting seasons and regulations as requested, while retaining responsibility for saltwater sport fisheries. The steelhead trout, for instance, is a freshwater game fish but, like the Atlantic salmon, makes nearly all its growth in salt water. It has not received adequate attention from either federal or provincial authorities until very recently. Even the traditional federal responsibility for research, both basic and applied, so well maintained in the fields of agriculture and commercial fisheries, has been largely neglected where sport fisheries are concerned.

These loose and flexible arrangements may have served a useful purpose in permitting sensible and workable patterns of responsibility to emerge. But the evidence now shows that the sport fishery is of direct concern to one in every three or four Canadians and of vital concern to the nation as a whole as well as to the individual provinces. It is a resource constantly threatened by other activities, such as logging and mining, and by increasing population, yet it becomes more and more essential as population grows and social stresses increase. In many parts of the country the sport fishery is an important area for direct or indirect investment and a valuable source of seasonal jobs. For all these reasons it rates full recognition as a major national resource, and requires the full effort of both federal and provincial governments to ensure that it is protected and developed to provide the best possible sustained yield. Yet it is not a resource quite like other resources in that its direct primary yield is pleasure and relaxation from participation in a human activity of respected tradition and almost universal appeal.

Chapter Two The Origins of Angling

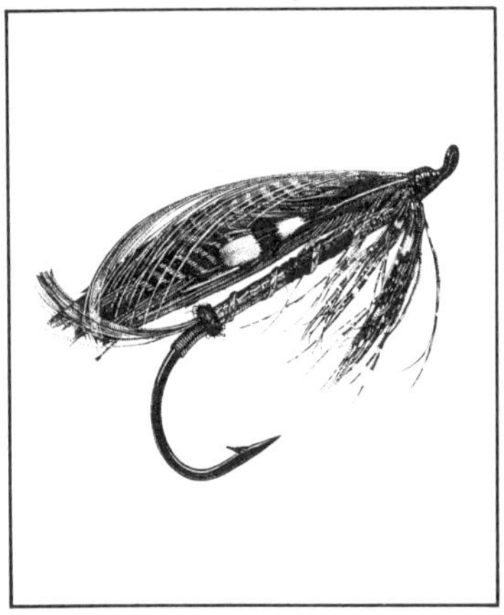

History does not record the name or the time of the first man who found pleasure in catching a fish; in all probability he was the first man who ever did catch one. From there to recognizing the pleasure in just going fishing may have been a long space in the story of mankind or a very short one. Man being man, the chances are it was very short.

Scholars have concluded that the first man caught the first fish with his bare hands. This great occasion probably happened as some big and barbelled fish struggled in muddy tropical shallows, the man came suddenly upon him and, almost without thinking, jumped in and took hold. I like to think of some later, but still very early, version when the man stood bearlike, thigh-deep in a clear and flowing northern stream as the silvery forebears of the salmons came surging up towards him. Then a single sweep of a pawlike hand would have pitched a fish on the bank to leap among the grasses or else searching hands would have found some larger quarry, seized it and clasped it against the hairy belly, while the fisherman struggled ashore. It is not difficult to suppose that on some bright and distant sunny day there were pleasurable sensations in all this, altogether apart from any ideas of food. Any ancestor of mine would have felt them, I am sure.

There was still to be a long progression to the art of angling or "fishing with an angle-hook." The scholars suppose that man moved first to the simplest projection of himself, the spear, hand-held and at first thrust and later, perhaps in some despairing motion of exasperation, successfully hurled after a disappearing target. After the spear, I think there would have been a trap of some kind, probably made by blocking off the exit of a tidal pool so that fish would be held there by the falling tide and easily caught. After that came the net in all its many forms.

The first baited line may have been just that, a line tied to a bait and hauled on when the fish had swallowed his end of it. The next step would have been the gorge, a stick or bone pointed at both ends, with a

Atlantic salmon *(Salmo salar)*
Rainbow trout *(Salmo gairdneri)*

Spearing salmon in the Bulkley Canyon, British Columbia

groove at the centre for the line. This device would have been buried in the bait and would have turned to prevent the fish from disgorging it when the fisherman hauled on the line.

The origin of the hook itself is lost in the dim mists of prehistory. The first hook may well have been a thorn or some appropriately shaped bone. William Radcliffe, the author of *Fishing from the Earliest Times,* insists that there is no evidence that the hook was used before Neolithic times, some ten thousand years ago; but thorns and bones would leave no evidence of the purpose they were put to. By 3200 B.C. the Egyptians were using well-made hooks of copper. By the year 2000 B.C. rod, line and hook were all in common use, and people fished for sport not only with rod and line but also with net and spear. This activity was certainly for sport, because ancient drawings of five thousand years ago show princes and other aristocrats at their fishing; it seems likely that the pleasures of tempting the unknown depths with hook and line reached into the heart of peasant's child and prince's son alike. Five thousand years of record and for how long before that? Far back into the ancient beginnings of human time, when the first man watched his line stirring, waited, then tightened to set the engorged hook.

The ancient Greeks went fishing with rod and line, Anthony and Cleopatra angled on the Nile, the walls of Pompeii have charming paintings of cupids and goddesses casting their lines along the rocky shorelines of the Mediterranean. Ælian gave us the first account of fly fishing nearly eighteen hundred years ago and it is clear from early paintings that the Chinese were fishing with line guides on their rods and modern-looking reels eight hundred years ago.

No doubt there was angling, both for food and pleasure, in western Europe throughout the Dark Ages. The first books dealing with sport are French, written in the thirteenth and fourteenth centuries and concerned mainly with hunting. Then, in the fifteenth century came Dame Juliana Berners's *Treatise of Fishing with an Angle,* the true forerunner of today's great wealth of angling literature. The *Treatise* is in no way fumbling or awkward. It celebrates and records a sport in advanced development, even to discussing natural insects and the dressings of a series of twelve flies which still catch fish today. It describes in detail how to make rods, lines and floats and offers an excellent examination of the ethics and aesthetics of the sport. All this was first printed in 1496, but was probably written a good deal earlier, and there may have been still earlier books that have not come down to us.

Over the next 150 years, until Izaak Walton's time, several other English angling books were published. There is ample evidence of the steadily increasing popularity of the sport: tackle was readily available and rods were sold in "every haberdasher's shop." Important new techniques were evolving, especially in the north, and the sport was ready for what perhaps was its first full flowering in any country. It was practised with enthusiasm by both cavaliers and puritans, and I am quite certain that it must have crossed the Atlantic with the early settlers. Nearly two hundred years separated Walton's *Compleat Angler* from the first North American angling writing of much significance, but it is natural enough that literature should follow somewhat behind action. There may also have been some measure of suppression; it has been suggested more than once that angling is a pastime for idlers and dreamers, and both the puritan ethic and the settlers' pragmatism may have tended towards this view in a land where the brook trout and the Atlantic salmon were quickly and easily caught in every stream. It seems significant that the Reverend George Bethune's first American edition of Walton, with its scholarly notes, appendices and exhaustive bibliography of angling works, first appeared anonymously in 1847 "owing to the public feeling against the propriety of such a book by a clergyman."

In 1849 *Frank Forester's Fish and Fishing of the United States and British Provinces of North America* was published. Forester's real name was Henry William Herbert. Herbert was an English professor who taught in New York and knew a very great deal about fish and fishing. He discusses Canadian salmon rivers and even notes that the rivers close to Quebec City have already "failed almost entirely," making necessary a sail of many hours down the St. Lawrence to ensure good sport. A second edition of the book, published in 1859, has a remarkably detailed account of the Pacific salmons, as well as "Gairdner's salmon," now known as the steelhead.

Just a year later, in 1860, there was published what seems to be the first Canadian fishing book: *Salmon Fishing in Canada,* by a Resident. The "Resident" in question was W.A. Adamson, and the editor was Col. Sir James Edward Alexander. In 1864 Thaddeus Norris published his much loved *American Angler's Book,* which included a substantial section on Canadian salmon rivers. Most of the rivers mentioned are along the North Shore of the St. Lawrence and most are still good salmon streams: the Goodbout (Godbout), "our best river"; the Moise (Moisie), "a very large river and full of fish"; the St. Margaret (Sainte-Marguerite) in the Saguenay, "it is over-fished"; La Val (Laval), "an excellent trout river"; the Natushquan (Natashquan), "a capital river, but very distant. I do not know anyone who has fished it."

It is obvious from these accounts that salmon fishing in Canadian rivers was a well-established sport. People came to Quebec's North Shore rivers from England and Scotland, from Maine and Massachusetts, from Montreal, Quebec and Saint John. Rivers were leased, guides, boats and canoes were available, and in those early days the sport fishery was a comfortable by-product of a lively, perhaps too lively, net fishery: in 1862 three rod fishermen, leasing the Moisie River for two weeks, caught 318 salmon, and the season's netting in tidal waters accounted for fifteen thousand.

It is natural enough that this early literature should have concentrated on the salmon, but most of the writers mention trout as well, both resident and sea-run. Special trout flies, such as the Montreal, were becoming widely known. Interest extended to the other side of the continent, in what had just become the Crown colony of British Columbia. His Honour Judge Matthew Baillie Begbie reported to the governor, from Queenborough (now New Westminster), that "two trout streams have been discovered: one running into the right of the bay opposite the shoal; the other nearly opposite Tree Island." These were the Brunette and the Coquitlam, neither in good shape today. But His Honour found many other good fishing streams and good hunting areas as he toured the country on his long and difficult circuits.

In January of 1881 the first issue of *The Canadian Sportsman and Naturalist* was published in Montreal. There seems to have been a number of very good naturalists in the country in those days and the

Lower Humber River, Newfoundland

Netted salmon from the St. Paul River, Quebec's North Shore
Salmon jumping in the Natashquan River, Quebec's North Shore

discussions in the various issues of the magazine are surprisingly up-to-date. In the January 1882 issue successive articles deal with a range of topics that might well find place in an outdoor magazine today. There is a complaint about "killing fish by dynamite," followed by a knowledgeable questioning of salmon hatcheries. This is followed by a piece in praise of field sports as opposed to athletics, then another that poses the question of whether fishing is childish or cruel.

Another magazine, *Rod and Gun in Canada,* started publication in 1899, and in 1901 reported the second annual meeting of the North American Fish and Game Protective Association, held in Montreal in January of that year. Representatives attended from Maine, Massachusetts, Michigan, New Brunswick, New Hampshire, New York, Nova Scotia, Ontario, Quebec and Vermont, and one of the objectives of the meeting was to harmonize the fish and game laws of the provinces and states. This seems to have been yet another sign of the developing maturity of approach to the management of the resource and to the sport itself.

Shortly after the turn of the century one comes upon native Canadian fishing writers. Napoléon Comeau and W. H. Blake were two enthusiasts for the North Shore of the St. Lawrence. Both are good writers and both share the ideals of the Canadian woodsman, Comeau as a lifetime professional, Blake as an amateur. Comeau was private guardian of the Godbout River for over sixty years and a federal fisheries officer for over forty years, but he was also a hunter, trapper, fisherman, self-taught doctor and midwife, postmaster, telegrapher, coroner and general servant of the people of the North Shore. His records of rod fishing on the Godbout go back to 1859, and he seems to have been a tireless salmon fisher himself; on 9 July 1874 he killed 57 salmon weighing 634 pounds, all on the fly. T. W. Lambert, who published *Fishing in British Columbia* in 1907, though he was not a native Canadian, based his book on twelve years' knowledge of the province. Lambert deals carefully, and for his time quite expertly, with the Thompson River, the trout lakes near Kamloops and the rainbow trout of the interior of the province, and his account of salmon fishing on the coast is a valuable record of the times.

While Comeau celebrates the Atlantic salmon and Lambert the

rainbow trout, Blake is a brook trout enthusiast. He sought them everywhere in the streams and lakes of the North Shore, "travelling light and faring hard" by canoe and on foot through the wildest country. The companionship of French-Canadian guides and encounters with the kindness, courtesy and good humour of the country people meant even more to him than the fishing, but he fished hard and thought nothing of two or three days tough travel to check on a rumour of "great trout" in some distant water. He must have been one of the first fishermen to travel extensively in the newly created Laurentides Park and he makes the firm claim for the park that "nowhere in the world do the *fontinalis* grow to a larger size than in these waters."

The early years of the twentieth century were a very active time in the development of the finer aspects of angling. In Britain F. M. Halford was perfecting the art of the orthodox dry fly, while G. E. M. Skues was developing the new art of nymph fishing. In North America Jordan and Evermann had just published the authoritative *American Food and Game Fishes,* Theodore Gordon was writing his wonderful letters to *The Fishing Gazette* in England about brook trout and rainbow trout in the streams of New York State and Pennsylvania, George La Branche was developing his ideas about fishing in his book, *The Dry Fly and Fast Water,* and there was steady interchange among these and other imaginative fishermen. The popularity of all kinds of angling throughout North America was about to explode.

Atlantic salmon from the Dartmouth River, Gaspé Pensinsula, Quebec

Hibb's Cove, Newfoundland
Bennett Lake, Yukon Territory

Chapter Three An Abundant Inheritance

Canada is variously estimated to have between a tenth and a quarter of the world's fresh water. For the most part this is cold water, so only a limited range of fish species is present, but the cold water species are among the world's finest game fishes. Thus the country has not only extensive freshwater fisheries, but a wide variety of high quality angling under every sort of circumstance and in every kind of scenery. In addition to fresh waters, the country borders on three oceans, Atlantic, Arctic and Pacific, having extensive shorelines and wide variations of climate. Sport fishing is a rich resource, so rich that Canada attracts more visiting freshwater anglers than any other country, and still can provide enough angling to satisfy her own people.

A sport fishery, like any other fishery, is based on three paramount factors: the natural fish stocks, the waters that nurse them and sustain them to maturity, and a favourable catching ground. It is true that fisheries of a sort can be maintained by continuous plantings of hatchery stocks, but they are so inferior and costly that they can be counted only as supplementary or diversionary, not as basic to the resource. Introduced species, such as the European brown trout and, perhaps less fortunately, the carp, must be counted as natural stocks wherever they have established self-sustaining populations. But it is a happy truth that Canada's sport fisheries are based mainly on the original native stocks. Atlantic salmon, five of the six Pacific salmons, two of the world's true trouts, four species of char, Arctic grayling, walleye, northern pike, muskellunge, smallmouth bass, yellow perch and several species of whitefish make up an impressive range of aboriginal stocks, all self-sustaining and all widely appreciated by anglers.

The waters that support these admirable stocks are, speaking generally and comparatively, still in good condition, capable of maintaining heavy annual yields and, where they have been damaged, open to rehabilitation. It would be misleading to suggest there has not been serious damage — in fact, only the immense size of the country and its relatively small population have moderated the overall effects of gross abuse. The shameful pollution of parts of the Great Lakes and of such great rivers as the lower St. Lawrence, the lower Fraser, the North Saskatchewan and the Saint John are well known. The Maritime provinces, especially New Brunswick, have suffered, and still suffer in reduced degree, from mono-crop forest practices and consequent disastrous spraying. There are many (and unnecessary) pollutions from pulp mills, mines, heavy metals of several kinds, various manufacturing processes, domestic sewage and storm drainage. All can be prevented and cleaned up, and one hopes they will be.

Badly conducted logging operations throughout most of the country have caused silting of streams and unnatural extremes of flow, from violent flood to summer drought. Careless agriculture through wasteful use of fertilizers has caused an excessive buildup of nutrients in the water, and even more serious damage has occurred from the unrestrained use of pesticides. Poorly planned or poorly managed irrigation has also caused much damage.

Perhaps dams have been the most destructive of the man-made evils inflicted so thoughtlessly on the natural worlds of fishes, especially anadromous fishes. The early settlers and loggers seem to have planted a small dam on every stream they came to, usually without a fish pass of any kind, or at best one that was inadequate. Usually there was a mill and more often than not some sort of pollution from its activities. Literally scores of fine local salmon stocks were wiped out in this way on both coasts, and the task of replacing them is slow and difficult. British Columbia has suffered heavily from dams on the Columbia River system, on several tributaries of the Fraser and on some of the Vancouver Island rivers. But so far, dams have not been permitted along the main stems of such major salmon producers as the Fraser, the Nass and the Skeena. It is devoutly to be hoped that they never will be.

The extent of past damage was clearly recognized by Scott and Crossman in gathering material for their comprehensive work, *The Freshwater Fishes of Canada*: "Time after time it would be found that a species had either disappeared from much of its former range (as is the case with the striped bass in the St. Lawrence, the whitefish and the lake trout in the Great Lakes and the Arctic char in Frobisher Bay) or that it had been completely extirpated from the American scene (as in the case of the blue walleye)."

But when all this is said and all the errors and losses are recognized, there still remains an immense resource, and much of the damage can, in time, be repaired. Pollutions can be cleaned up and brought under full control; small dams can be removed or by-passed. Watersheds devastated by fire or logging or both renew their ground cover. Damage to estuaries and wet lands is perhaps more serious than any other, but the true importance and productive value of these lands is more generally understood than it used to be and there is now a significant effort to protect them. In some instances I believe rehabilitation will be possible and in others perhaps positive improvement.

There has been a tendency in the past to underestimate the importance of what I have called "the catching grounds," yet these are essential to the resource. All fishermen, whether they realize it or not — and most do — care very much about the aesthetics of their fishing spots; these are all part of the mood and meaning of going fishing. Aesthetic qualities are rather easily destroyed. But there are pragmatic values also — in free-flowing streams, in shallow flats in freshwater lakes or along saltwater shorelines, in river mouths, over reefs, around islands. It is important to recognize such places and to protect them from excessive water traffic and industrial, residential or recreational encroachments. Once the principle is understood and accepted, alternate sites for these other uses can usually be found without much difficulty.

Canada still has hundreds of thousands of lakes, thousands of streams and hundreds of miles of sheltered shoreline where exceptional beauty can be found and good stocks of fine fish as well. The provinces present a splendid parade of fishes and fishing waters, from east to west, from the southern boundary to the Arctic. There are wild waters and wilderness waters still, waters hidden in the bush or accessible only by foot trail, waters that can be reached along difficult tracks or abandoned logging or mining roads, waters in parks, waters along highways and byways, crowded waters and waters where an angler can still find solitude.

The parade of fishes, province by province, is just as impressive, though in some ways deceptive. In the Maritimes and eastern Quebec, the Atlantic salmon is famous. A noble fish, certainly, of noble habits and the peak of ambition for most freshwater fishermen. But the fisherman's fish of the Maritimes, overwhelmingly popular, is the brook trout, the eastern speckled trout *(Salvelinus fontinalis);* in Quebec the speckled trout, *la truite mouchetée,* is "the fish of the province" and the province may well be considered the centre of its abundance.

Newfoundland, Nova Scotia, Prince Edward Island and New Brunswick all play host to the bluefin tuna *(Thunnus thynnus)* in late summer and early fall, a seagoing monster that reaches weights of well over a thousand pounds. The bluefin is the only "big game" fish that Canada has to offer, but he attracts fishermen from all over the world to the international tournaments. The Arctic char *(Salvelinus alpinus),* handsome and a little mysterious still, is proudly claimed by Labrador and the Northwest Territories, but also by Newfoundland and Quebec.

The lake trout *(Salvelinus namaycush),* the lake whitefish *(Coregonus clupeaformis)* and the northern pike *(Esox lucius)* are the most widely distributed sport fish in Canada. The lake trout is present all across the country, from Nova Scotia to the Yukon and northern British Columbia and high into the Arctic, except in the immediate drainage around Hudson Bay. The northern pike does not reach quite so high into the Arctic, but is found in every province from Quebec to northeastern British Columbia and the Yukon and through most of the Northwest Territories, including the Hudson Bay drainage. The whitefish's range is almost the same except that it is found in New Brunswick and well into British Columbia. Lake trout and lake whitefish are very important game fish in Quebec, Ontario, the prairie provinces and throughout the north. Whitefish are extensively taken by ice fishermen, but the possibilities of this fish have not yet been fully discovered by summer anglers. All three species are of great commercial importance.

Both smallmouth bass *(Micropterus dolomieui)* and largemouth bass *(Micropterus salmoides)* are very popular fish in southern Ontario, and the muskellunge *(Esox masquinongy)* is a highly regarded trophy fish because of his potential size and formidable fighting reputation. But in terms of numbers caught by anglers the most important fish is the

Tuna tournament, Caraquet, New Brunswick

Walleye *(Stizostedion vitreum)*
Yellow perch *(Perca flavescens)*

yellow perch *(Perca flavescens),* closely followed by the walleye or yellow pickerel *(Stizostedion vitreum).* Both these fish are excellent on the table and both are caught under the ice in winter; the yellow perch is also much used as a bait fish, since there is no size limit. The popularity of the walleye extends beyond Ontario eastward into Quebec and it is perhaps the most keenly sought of all sport fish from there to the Rocky Mountains. It happens that the walleye is not the most easily found or most easily caught of all fish, so many walleye enthusiasts end up with northern pike instead; but at least the intent was there and probably will still persist on future occasions.

In the prairie provinces, after the walleye, the important natives are northern pike and lake trout. Manitoba has some smallmouth bass in the south. In Alberta rainbow, cutthroat and brook trout thrive in streams and potholes along the Rocky Mountain slopes. In the northern parts of these provinces, on the Arctic watershed, *le poisson bleu,* the Arctic grayling *(Thymallus arcticus),* is always ready to rise to a small fly. Grayling are abundant in the Yukon and the Northwest Territories. The Yukon matches the Arctic char of the Territories with its very close relative, the Dolly Varden *(Salvelinus malma).*

On the western slope of the Rockies, all is different. British Columbia is the home of five species of Pacific salmon, the cutthroat trout *(Salmo clarki clarki)* and the rainbow trout, *(S. gairdneri).* Most anglers are trout or salmon fishermen, though such fish as the mountain whitefish *(Prosopium williamsoni)* and Dolly Varden play a part, and there are grayling in some of the northern watersheds. Of the Pacific salmons the chinook *(Oncorhynchus tshawytscha)* and the coho *(O. kisutch)* are considered game fish and provide most of the sport in both fresh and salt water, but in years when big numbers of pink salmon *(O. gorbuscha)* are plentiful, anglers catch a lot of this species also. The sockeye *(O. nerka)* and the chum *(O. keta)* are much less often taken by anglers, though this may change. The kokanee salmon, which is a freshwater resident sockeye, is extremely abundant in some of the big lakes in the province, such as Shuswap and Kootenay, and very important to anglers. Most kokanees mature, spawn and die as four-year-olds of ten or twelve inches in length, but the stocks in Kootenay Lake, especially in the west arm, grow much larger. Fish of seven or eight pounds are not unknown and two- to four-pounders are quite common. The marked increase in growth rate coincided with the successful introduction of a small shrimp, *Mysis relicta,* to the Kootenay system.

Both rainbow and cutthroat trout have sea-run forms as well as local stocks that differ widely enough to call for different popular names, even though the physiological differences do not justify even subspecific distinction. The rainbow is so highly adaptable (and has been so extensively planted) that it is becoming an important sport fish in every province in Canada. In B.C. the seagoing stocks have become famous as the steelhead, the Pacific Coast counterpart of the Atlantic salmon. The interior stocks, in lakes usually above the 2,000-foot level, are known as Kamloops trout. Resident stream or lake stocks at lower elevations are usually called rainbows. The cutthroat clings to his original name but with modifications that are readily expressive to the angler: sea-run cutthroat; coastal cutthroat; and, in the Kootenays, Yellowstone or black-spotted cutthroat.

Because they live along the edge of climatic change, the trouts and chars and salmons are nervous creatures. Only a few degrees of temperature change would enormously restrict or expand their range and they seem always ready for it. Rainbow, cutthroat and brown trout all have sea-run types, as do brook trout, Arctic char, Dolly Varden and, to a more limited extent, lake trout. The Atlantic salmon has its landlocked forms — Sebago and ouananiche — and at least two Pacific salmon, the sockeye and the Japanese cherry salmon *(O. masou),* have resident freshwater forms. All the Pacific salmons can be raised to maturity in fresh water.

This then, in broad outline, is Canada's parade of game fishes, and it is an impressive one in terms of quality and for the most part the health and strength of the stocks. The resource, considered in all its aspects of habitat, stocks and beauty of the surroundings, including the wealth of other wildlife, the wide variation of terrain and flora, is clearly outstanding. Given adequate research, sophisticated management and the solid co-operation and understanding of both anglers and resort owners, it can continue to provide large numbers of fishermen with the highest quality of recreation indefinitely.

Smallmouth bass *(Micropterus dolomieui)*

Rocky Mountain whitefish *(Prosopium williamsoni)*

Interior lake, British Columbia

Chapter Four The Pleasures of the Sport

Going fishing means different things to different people. To me it means putting on a pair of waders, picking up a fly rod and going off to a stream to look for fish. I prefer to fish alone or with one or two companions who think and fish pretty much as I do. I agree that I would still be going fishing if I went to a lake or to salt water and climbed into a boat with my fly rod or, under some circumstances, a bait-casting or trolling outfit. I should certainly be going fishing if I were wading the ocean surf and casting a lure for striped bass or bluefish.

These are just personal preferences that come readily to mind, but it is important to realize that all fishermen have such preferences and that they may differ very widely. The most superficial observation of sport fishing, or for that matter of other forms of outdoor recreation, reveals two types of approach: the solitary and the gregarious. Extreme examples would be the wilderness backpacker or canoeist and the camper who happily crowds his motorized home-away-from-home into the confines of campsite or trailer-park between neighbours so closely placed that he could reach out and touch them. Both want the wilderness experience, but the former wants to face it head on as an intense and deeply personal affair, and the latter prefers it at a safe distance, kept so by surroundings and companionship much like those of his crowded and comfortable daily life. What is the measure of the two experiences? It is easy to judge which is *likely* to be the more intense, the more rewarding, but it is important to remember that they are being enjoyed by two quite different people. Who is to say that the slant of sunbeams through massive forest trunks is more noble and inspiring in wilderness solitude than in some roadside park? Much depends on what the viewer brings to the scene. The one certainty is that the experience of the crowded campground would have very little meaning without the existence of the true wilderness experience. Somehow a measure of the quality and mystique of the full experience is transferred to the other.

Falcon Lake, Manitoba
Maligne River, Alberta

Much the same is true of angling. The very highest and most sophisticated forms of the sport provide the strongest and deepest satisfactions. But it is the practice of those highly developed forms that reflects back on the whole of angling and helps to give it breadth and depth of appeal. To breadth and depth should be added variety of appeal, and it is important to examine in some detail the nature and scope of this variety.

There is not one way of going fishing, but many, and each has special appeal and meaning for its devotees. It has been a long time since I have done it, but I can recall very clearly the intense pleasure of watching a bobber with a worm or a minnow on a hook somewhere in the water underneath it, the mounting excitement that came with the first signs of something going on down below, the firm drawing of the bobber along the surface of the water and its sudden slanting plunge below. This way of fishing is usually considered rather unsophisticated, though it is capable of great refinement. Most fishermen, in spite of their reputation for patience, are inclined to be somewhat restless and probably prefer the active searching of water by casting, or even by trolling, a lure or bait or fly. Many fishermen dislike fishing with bait and some prefer to limit themselves to nothing but a fly. Many of us find our chief pleasure in some single moment of fishing, perhaps the slow visible rise of a big fish to a fly drifting on the surface or fished just under the surface film. Some of us get more satisfaction from watching two otters playing in the pool we are fishing or an osprey hunting over the shallows than from catching many fish.

The interesting thing is that these variations on the theme of going fishing, and many more besides, can all be the preference of one man at different stages of his life. Equally, they can be the confirmed preferences of various men throughout their lives, though this is not likely. At twenty I delighted in wading swift rocky streams in jeans and old logging boots; at thirty I waded the same streams, enjoying the same challenge, but wearing waders and using considerably more skill and finesse; at forty and fifty I was no longer looking for challenge, but I was strong, well balanced and confident. Today, in my late sixties, I go no deeper than I have to and use my head to avoid any extreme stress on the uncertain legs. But I still enjoy the feel of strong water about me and the calm deliberation of finding a secure foothold and advancing firmly upon it.

Wading is one small aspect of fishing, perhaps of fly fishing especially, because the reach of a fly rod is more limited than that of a bait or spinning rod and far more limited than that of a man in a boat. Casting is a sharper and more essential pleasure in going fishing. Casting a fly with precision, comfort and control is an intensely satisfying exercise; so also are the skills that set spoon or spinner exactly beside a chosen rock or lily pad a hundred feet away. There can be a fine and sometimes critical exercise of skill in controlling a floating fly so that it drifts freely without any pull of the line to drag it unnaturally across the flow or in working a sunk fly at some precisely chosen depth and speed. The same skills and satisfactions are in the working of lure or bait to greatest effect. A host of sensitivities is involved, of hand and mind and eye: reading of the stream surface; measuring of speed against slack or eddy or current flow; knowledge of bottom contours or obstructions; above all, recollection of past experience, that tests what is happening under the water.

All these things are pleasures of performance, yet without reasonable possibility of some response from the fish they would not count for much nor last for long. There is always, or nearly always, the preferred fish, but some other may be an adequate substitute either as a novelty or a welcome alternative. The yellow perch is not very large nor a spectacular fighter, but Ontario anglers catch nearly thirty million a year. The yellow perch is a very delicate biter and it takes skill to hook him; he is also a very good eating fish.

This perch — along with walleyes, whitefish and lake trout — is a favourite of ice fishermen, and ice fishing is a highly social form of the sport: people set out huts, cabins, villages and small towns on the ice, visit back and forth, play cards, drink a little, catch a few fish. But there is also a solitary form of ice fishing. In Regina I talked with a husky young man who was poetic in his enthusiasm: "In the winter I often leave Regina at midnight and drive three hundred and fifty miles north to my favourite lakes. I spend two days and two nights fishing there, sometimes alone, sometimes with friends or just the family. The north

Salmon fishing, Nova Scotia

Ice-fishing catch, Lake Simcoe, Ontario
Rainbow trout in Dragon Lake, British Columbia

is so beautiful, so different in winter. People think it's just a big snowbank, but it isn't. It's so peaceful there." This man and his friends often spend their time watching fish under the ice rather than fishing for them. The procedure is to rig a black cloth, much like an old-fashioned photographer's hood, to shut out overhead light and simply wait to see what happens down near the bottom of the lake. I understand that this form of fish watching is becoming increasingly popular in other provinces as well.

In any fishing the response of the fish to lure or fly or bait is obviously a high point, and the angler's reaction is usually crucial: if too slow or too fast the fish will not be hooked, if too heavy the leader may break, and if too light the hook will not be driven home. A big, hungry northern pike or black bass will likely hook himself in the predatory determination of his attack. But a big brown trout rising to a floating fly calls for very precise timing, and the slow, deliberate take of an Atlantic salmon or steelhead in low water calls for firm control. The general rule is that the bigger the fish the slower must be the angler's response in raising the rod to set the hook.

The culmination of this series of performances and occurrences is usually considered to be the "fight." Some fish are fast, active and violent and some are not. Conditions for controlling and tiring the fish may be favourable or unfavourable, but I think that most of us who have fished a good deal feel that the odds are with us once the fish is securely hooked. Even so, fish can do unexpected things; a long fast run, a series of wild jumps, or a strong drive towards an obstruction can be exciting if the tackle is properly geared to the size and strength of the fish. When the fish is played out, there is still another choice: one can kill the fish and eat him or release him from the hook to swim away and do it all again for someone else.

These are the bare bones of the angler's experience by the waterside and no more than that. Put them all together and they describe the act of fishing for and catching a fish; but a great deal more than this is going on. First of all, the angler has probably spent some hours or even days in preparing his gear and planning the details of his trip. He gets himself outdoors in either new or familiar surroundings and is suddenly aware of many things that are foreign to his daily routines. The air is fresh, the surface of the water has patterns and colours. Forget-me-nots, erythronium lilies, prairie crocuses or some other flowers of the season are in bloom along the banks. There is sun or wind or rain or snow. The river is high, the lake level is low; leaves are opening, leaves are falling; birds are moving, lively in spring, quiet through the nesting season, glad with their broods later on. There are flies — stone flies, mayflies, damsel flies, dragonflies and probably biting flies as well. How much of this activity each individual angler takes into his consciousness no one can really know, but it is the setting of the sport and inevitably important. Generally anglers have noticed it all with some thoroughness ever since Dame Juliana explained that even if an angler found no fish he still had his "holsom walke" and the "swete ayre of the swete savoure of the meade floures — He hereth the melodyous armony of fowles." If, in addition, "the angler take fysshe: surely thenne is there noo man merier than he is in his spyryte."

Even away from the water, the meadows, the flowers and the fish, most anglers are likely to gain a lot of satisfaction from their sport. I have already mentioned the joys of preparation and anticipation. Anglers tend to be readers, whether they enjoy the gentle bucolics of Izaak Walton or the awkward *machismo* of the battles with "lunkers" described in the popular American sporting magazines. Nearly all anglers fuss lovingly over their tackle at some time during the closed season.

Most serious fishermen develop some fairly intense line of specialization within the sport sooner or later and follow it to considerable lengths. Fly fishermen are likely to become fly-tiers and often good stream and lake entomologists as well. Many fishermen make their own rods, sometimes with a craftsmanship beyond price. Quite a few are ingenious inventors and makers of lures; some are boat enthusiasts, some become camera enthusiasts, some collect angling books, some compete in casting tournaments — the possibilities for specialization are almost unlimited and the hours spent upon its many forms are beyond count. I fuss over tackle a little, tie flies occasionally, practise casting on dry land, but my form of specialization has always been the fish themselves, their life histories and habits and anything else I can learn about them.

The point is that all this pleasure and interest could not exist without the resource — the fish and the water. The depth and quality of the sport and the rich nature of its appeal cannot be overemphasized, nor can the obligation the state has to serve these more serious and dedicated anglers well. In many ways these anglers make the sport, preserve its traditions, build on them, and give real meaning to the whole affair. It is true also that fishermen of this calibre are fulfilling a duty that all fishermen have to the resource and to the state — that of realizing the greatest possible pleasure and satisfaction from every fish they catch.

Chapter Five Why Go Fishing?

Perhaps the question "why go fishing" was answered to a degree in the last chapter, but there is still a good deal more to say about it. There are endless surveys of the matter, leading to endless statistics, which in turn often lead to the remarkable discovery that people go fishing to catch fish. It is fairly obvious that everyone goes fishing with the general idea of catching fish, otherwise one would be doing something else, like bird watching or botanizing or snorkelling or simply going out for a "holsom walke." But if the idea really is to catch fish, angling isn't much of a way to do it — gill nets, trap nets, seine nets, set lines, spears and dynamite are all more efficient under most circumstances.

Similarly, some surveys discover that people go fishing because they like to eat fresh fish. Certainly this is one way of making sure the fish is fresh, if it hasn't been left too long in the sun with the flies, but it is a very expensive way, and I know lots of anglers who don't particularly like eating fish at all. Fresh walleye fillets by the waterside are fine, but *filet de doré* in a Montreal restaurant is likely to be much better. The idea that "everything tastes better outdoors" really means that one shows a much greater tolerance for indifferent cooking when out in the open air.

This criticism is unfair, of course. There have to be facts and figures to show the use of the resource, the public interest in it and demand for it, the dollars it shifts from one pocket to another, the jobs it provides and the tourists it attracts, if politicians and public are to be persuaded to treat the resource seriously and provide properly for it. Questionnaires have to find a consensus of some sort and questions have to be designed in broad terms to provide for this. Questions such as How many days did you fish last year? In what months? What fish do you prefer to fish for? can be answered and are likely to be answered with fair accuracy. But the question Why do you go fishing? is much more difficult.

Fortunately, the questions actually asked are a good deal more subtle and sophisticated, though they usually call for multiple-choice

Bluefin fishing, Nova Scotia

Miramichi River, New Brunswick

answers that put words in the mouth of the witness. The answers also depend a good deal on the nature of the sampling. Two different groups of saltwater salmon anglers in the Strait of Georgia gave somewhat different ratings to questions asked in recent surveys. One group gave as the preferred reason for going fishing: appreciating the thrill of catching a fish; the other group felt that the experience of the catch rated no better than fifth. Other factors rated in the first five reasons by both groups were: being outdoors; taking life easy and getting away from working pressures; the companionship of friends; observation of nature. One group rated eating fresh fish third on the list, but the second group was not given this choice.

Again, this brief account is quite unfair because both surveys asked many more questions and examined a whole assortment of issues important to fisheries managers. I offer it simply to indicate the difficulty of finding simple answers to questions that involve highly complex and variable human factors.

Both game and fishery managers have, until recently, been far too concerned with the "harvest" from the resource — the number of man-days of hunting or fishing produced and the "success rate" per man-hour of fishing or hunting. These statistics miss the point entirely, because they fail to measure the quality of the experience realized.

Another difficulty with surveys is that they probably tend to reflect the casual or marginal angler too heavily — that is, the man, woman or child who fishes because it seems to be a part of a holiday. This individual and his or her preferences are undoubtedly important to the tourist industry, but they have no validity whatsoever in terms of what sport fishing is all about. The marginal fisherman may be going in either of two directions — right out of the sport altogether or more deeply into it. If the former, there is not much need to worry about him — some other form of recreation will probably claim his attention. But if it is the latter, and especially if he is young, his preferences should be carefully examined, because angling needs likely recruits in every generation and there can be no doubt that a deep interest in angling will be of value to him all through his life.

I was interested to hear the other day, as a result of a survey, that it takes an average of .12 fish per hour (or a fish every day and a half fished) to keep an Atlantic salmon fisherman happily fishing away. At first glance this seems like a very sound figure since both Atlantic salmon fishermen and steelhead fishermen have long known that an average of one fish a day is enough to keep them going on ordinary public waters. But I have been worrying ever since about whose day and whose hour was being considered. A "day" is not likely to mean eight hours of steady fishing if one allows time out for changing flies, for lunch, for resting, for chatting with a companion and so on. An "hour" might well mean sixty minutes of pretty steady fishing, but even here there will be time lost in changing pools; or if it is all spent in the same pool the fisherman is pretty sure to be fishing for twenty or thirty minutes of the time over water where he couldn't possibly expect to catch a fish because it is convenient to do so and just in case there might be something there. Fishing and fishermen are so much more interesting than statistics.

This brings up the question of how far management and regulations can pander to mass preferences without destroying the sport. In spite of the available tradition from Europe, angling as a sport got off to a fairly bad start on the North American continent. Early angling was primarily food fishing, just as early hunting was primarily meat hunting. There is nothing bad about that in a new country where the need exists, people are few, and fish and game are abundant. But the abundance itself was a trap, leading to the pathetic concept of the "inexhaustible resource" and to generations of fish hogs and game hogs who demonstrated their prowess by having their pictures taken in front of serried ranks of slaughtered fish or game. Napoléon Comeau's accounts of early catches (including his own) on the Godbout and other rivers reflect an obsession with numbers that would have left little time for relaxation or contemplation. He writes almost pityingly of a John Manuel, whose skill he greatly admired, that his two best days, in what was a poor year, "yielded but 8 and 7 fish respectively to his well-handled rod." Really good days produced catches of twenty to thirty fish. W. H. Blake, writing of the last quarter of the nineteenth century, offers this: "Every stream abounded in trout. The reward for a day of good hard work . . . would be from three to six dozen trout worthy of being basketed (I remember 24 dozen for two rods and a half)."

Neither Blake nor Comeau were "fish hogs" themselves. Comeau was a professional and responsible for the catch on the Godbout, and Blake had written of the Laurentides Park lakes: "It is no extraordinary feat to take five or six dozen trout in an hour, but it is to be hoped that a very few experiences of this kind will satisfy." He himself seems to have been endlessly in pursuit of few and very large brook trout. But the attitudes of both men express very clearly the great numbers of fish that were available and the all too frequent response of anglers to this abundance. Except in the case of Atlantic salmon streams, practically all this fishing, from one side of the country to the other, was free and open for the taking, without so much as the cost of a licence.

So Canadian anglers have in their heritage three unfortunate concepts: there is an unlimited supply of fish; regulations are unnecessary, and fishing should be free. These concepts must be eliminated. The resource cannot support a meat fishery, it cannot be enjoyed for long without close regulation and management, and if it is to be managed properly, there must be a source of revenue — it cannot and should not be offered without direct cost to the participants in the form of licences.

All this regulation is gain rather than loss. The resource is put in proper perspective, as something of immense value, to be cherished, used respectfully and passed on unimpaired to future generations. True, management also brings about some loss of wild freedoms and in some sense a loss of quality; but these are penalties of increasing population. Angling, if it is to persist, can only do so as a sport of high principles, strong ethics and intelligent recognition of the true nature of the resource. Such principles, like the ordinary concrete regulations that bind him under the law, are not a burden upon the angler but positive enhancements of his chosen pursuit. To be fit to make proper use of the fishery, he has to bring something more with him than a rod, a line, a hook and a desire to kill fish.

This is not a restriction, but a challenge — the same sort of challenge that human beings meet with consciously or subconsciously in everything they undertake: How can I do this thing that I have to do (or want to do) so that it has the most meaning for me and for others? This feeling does not always have the same intensity, although it is the foundation of a lively life. Is the question perhaps too heavy to be applied to this quiet and gentle invasion of the lakes and streams and

Algonquin Park, Ontario

A day's catch, Prince Edward Island

countryside? I am sure not, if only because it is no small thing to inflict oneself on the private world of fishes, even if one does not kill. To go beyond that and take a life or several lives should always be a matter of consideration for a thinking person.

In extreme youth one probably has more concern for the worm impaled on the hook than for the fish triumphantly hooked and landed. The mature fisherman, too, is likely to answer the philosophical question "why go fishing" with his senses rather than with positive thought: I, this human entity, gain immensely more than any loss I am inflicting. The fisherman who articulates an answer to the question may add: I am using the resource carefully, well within its capacity to maintain itself. I am learning more about it and its meaning. Without my concern and interest, self-centred though they be, in all probability the resource would be entirely destroyed by some other human activity.

If these are rationalizations they are an improvement on the simple biblical concept that God gave mankind "dominion over the fish of the sea, over the fowl of the air, and over the cattle, and over all the earth, and over every creeping thing that creepeth upon the earth" — a concept that more nearly reflects man's ruthless abuse of the natural world than an abiding respect for it. But the real justification of the sport is in the focus it provides for intense human interest in the fish and the waters and everything that lives and moves on or about them, and the vital, luminous experiences that develop from such interest.

Are all forms of angling experience equally significant, important, valuable, intense, rewarding to the participant? For one man, fishing may be a lifelong interest. Everything he does may be in some way built around it — his job, where he lives, his close friends, what he reads, what he dreams. For another, going fishing may be a brief fad or fashion, a means of advancing a business deal or simply of getting out of the house. Obviously the two will bring totally different levels of performance and appreciation to the sport and will experience entirely different sensations and satisfactions. For the lifelong fisherman the day, whether it was good or bad, will have added something to the pattern of his life, perhaps even yielded some ecstatic moment to be remembered and cherished forever. To the casual fisherman the day may have yielded much or little, but it is quite unlikely that its fishing aspects will have seemed very rewarding or be long remembered.

These are two extremes that may differ as widely as the ecstasy of the violin virtuoso in an outstanding performance differs from the frustrations of the beginner at his exercises. In between the two extremes there is an infinity of variations, but the difference between the confirmed and the casual angler should always be apparent. I believe it is a difference of great importance to fisheries managers, because the dedicated angler will almost certainly be one who contributes to the sport as well as takes from it. The highly diffuse, amorphous consensus produced by such individuals develops the techniques, standards and ethics that give continuous life and meaning to angling. There are clubs and associations that often combine into provincial, national and even international federations, and these groups discuss, examine and may adopt what comes forward to them from the creative level of the individual angler. But it is still the individual by the waterside or the few friends who meet to talk fish and fishing over a drink or two on winter evenings that make angling what it is, a major human occupation of almost universal appeal and one that continues down the centuries.

Cormorant Lake, Manitoba

Releasing a steelhead into the Englishman River, British Columbia

Chapter Six Guarding the Standards

It is quite easy to debase the sport, change its values, dilute its ethics and destroy its traditional associations with quietness, relaxation and the opportunity to think. Angling is not a competitive sport. The fisherman's only real competition is with his quarry and his only real challenge is the challenge to himself. Nothing can add to this, but the blight of interhuman competition can certainly detract from it. There are casual forms of competition that are relatively harmless, such as a side bet between friends or the affectionate regard for fish of record size taken on tackle of closely defined specifications or the many types of local club awards that have more to do with good fellowship than competition. But there is little good to be said for the so-called fish derbies, which are nearly always directed to commercial ends by tourism entrepreneurs and advertisers of all kinds. At best they are a perversion of the real nature of the sport; at worst they can be seriously damaging to the resource itself.

Sooner or later competitions inevitably concentrate too much gear on limited quantities of fish. The larger the prizes, the greater the concentration will be, and year by year the number of fishermen increases even without the stimulus of increasing prizes, so competitions that may be safe enough today can be confidently expected to cause damage before very long. In other words, derbies are bad in principle and bad in practice. Few biologists question the fact that the stock of big lake trout in Amisk (Beaver) Lake in northern Saskatchewan was wiped out by local trout derbies. Although not favourable to lake trout reproduction, the lake carried naturally a stock of very large fish, but the stock could not sustain itself against concentrated pressure. Closely managed, it could have been maintained indefinitely.

Put-and-take fisheries are a betrayal of the essence of the sport except in waters incapable of maintaining a natural stock, such as prairie ponds and dugouts and lakes with inadequate natural spawning areas. Put-and-take stocking can also be justified in waters close to population centres, but even here it should be carefully planned to preserve as much as possible of the atmosphere of a wild fishery. Anything that encourages a chase-the-hatchery-truck mentality in young or old is bad.

One of the worst examples of debasement of the sport I know of was tried repeatedly in the state of Washington, where "snagging seasons" were opened and "anglers" encouraged to catch excess stocks of salmon returning to the hatcheries. I happened to notice and keep two Washington announcements of the same date, 16 September 1974. The first set the opening of a "special snag fishery in the Washougal River" on 19 September. The daily bag limit was six salmon. The second announced a closure of the sport fishery in the Kalama River on 18 September "due to the extensive unlawful snagging of chinook salmon." It adds: "Reports indicate that as many as 200–300 fish per day are being deliberately snagged." The two announcements suggest rather strongly that you cannot promote a rotten ethic and expect to escape the consequences. But that is not really my point. Even though it is done with hook and line, deliberate snagging is not fair angling. Any attempt to make it so by legislation is an attempt to destroy what the sport is all about.

Hatchery stocks can force managers into still other aberrations, such as trucking anadromous fish back to the mouth of the stream to run them three or four times through the fishery. The problem is that hatchery stocks are bred so that the fish all run through within a season of a few weeks instead of four to six months. Since it costs ten dollars or more to bring back an adult steelhead even once, they must be quite expensive fish by the time they make the third or fourth trip. It would seem also that they must be pretty mature and just a little tired in spite of the free rides downstream.

Extreme mechanization of the sport is also debasing, though there is some merit in the fact that it extends the range of anglers and so reduces crowding. But mechanization needs control and too many of the provinces have been slow about it. All impending abuses call for early and prompt control, because it is so much easier at that stage than when the abuse becomes entrenched. There is still far too little regulation of the size and use of outboard motors on lakes and rivers in most, if not all, provinces. All-terrain and four-wheel drive vehicles, dune buggies and bush bikes are out of control from Newfoundland to

British Columbia and are responsible for untold damage to soil and vegetation throughout the country. It should be clear to any angler, or any outdoorsman for that matter, that extreme mechanization is indefensible on environmental grounds alone. Anglers, above all people, need clean water, clean air, good earth conditions, good sea conditions. The power-driven boat, the all-terrain vehicle, the large recreational vehicle all work against the environment, not only in their direct use of energy but also in the waste of energy and materials in their manufacture.

In terms of overall management there is a good case to be made for simplicity of gear and equipment. It can help to maintain the aesthetic and spiritual qualities of the resource and at the same time protect the physical values. I realize that many anglers get much of their enjoyment from complex and ingenious equipment, and that affection for internal combustion engines is still almost unbounded; but a little restraint, voluntary or imposed, will pay off handsomely for anyone who sets much store in going fishing. Unfortunately commercial interests are all on the other side and have far too much to say about how we run all our recreational resources.

It seems important to recognize that the recreational fisheries resource has two essential aspects. The first is the resource itself — the fish, the waters and their surroundings. The second is the sport, without which the resource could not be said to exist. The physical components of the resource quite clearly belong to the Canadian people; but the sport — the traditions, standards, ideas and ideals that make angling a sophisticated and valuable recreation — belong to the anglers. The anglers are not of one generation, but many; they are not of one nation only, but of all nations that share the history and traditions and practice of the sport. Sport fishing is international and even though there may be national, provincial and local differences, the basic concepts are the same.

The true sportsman, whether hunter or fisher, has always had a certain sense of intimacy and identification with his quarry. There is an obligation to know and understand the animals on water or land, to understand their needs and respect them, to respect the breeding season, to respect and admire the strength and endurance and persistence of the animals in their migrations and in their will to survive. These are religious or semireligious concepts, built into the heart of man aeons before there was sport as such, reawakened and cherished in the heart of the modern sportsman because he is in search of his own place in the wild, his own identity in the natural world.

The search may be tough and aggressive or gentle and sensitive. Among fishermen, guided whether they know it or not by the gentle precepts of *The Compleat Angler,* it is usually the latter. The word sport, in any connotation, implies a sense of generosity towards the opponent, a desire to meet and test honourably under conditions fair enough to ensure that the outcome is uncertain. If these implications are lost or destroyed, the meaning of sport disappears. In athletic sports the ideal is often lost for commercial or competitive reasons. In field sports, such as fishing, there is no sense or reason in losing the ideal, since the sportsman is at most testing himself and that not too earnestly if he understands his sport properly. He is in the world of the river or the lake or the sea; he is part of that world because of his concentration. He is intensely aware, watching for signs, for movement, for anything that will help his quest — and, if he is wise, for many things that will not help it at all, but will enrich his day.

The sense of generosity towards and affection for the quarry that I have mentioned so casually is a very complete thing that finds expression in many different ways. In the present context it means, for instance, that the angler will select gear that is reasonably matched to the strength and speed of the fish, that he will choose the method of fishing that offers the greatest challenge with a reasonable chance of success. It means that he will respect the fish's world, moving through it with the least possible disturbance, leaving only the faintest possible trace of his passing. It means respect for the fish's world in another sense, in that he will give both thought and effort to the preservation of that world, that he will protect and affectionately release undersized or unwanted fish.

In the faithful observance of these duties the other dimension is realized: the sense of belonging there on the lake shore, at the river's edge, along the kelp bed; the sense of participation in something more than a simple attempt to catch a fish. One is no longer an intruder. More

Salmon fishing, Nova Scotia

Little Pine Lake, Ontario
Maligne Lake, Alberta

than once, wading or swimming among them, I have told the nervous spawning salmon: "Relax. I belong here as much as you. I have done my best to understand. I have tried to help. I have earned a place." If this sounds an unlikely monologue, I can only give assurance that it comes out very naturally over the sound of the swift water. I don't suggest that the fish make very much of it, though I try to guide and control my movements to express reassurance.

I wrote in another book of Long Baptiste, the great Dene hunter who guided the Dunleavy and Sellers outfit to a rich strike on the Horsefly River during the B.C. gold rush in the mid-nineteenth century and later guided Judge Matthew Begbie to both hunting and fishing. The Indian was a big man, a superb athlete, a jovial companion, a very quick mind, and his sense of identity with woods and wildlife was complete. One morning, as they were breakfasting on a deer he had killed before they were awake, he scolded the Dunleavy party for throwing the bones in the campfire: "Not burn good bone," he said. "Coyote all time behind where man go. Hungry, find bone, eat. Tell Indian thank you." Then he howled like a coyote and the answers came from all about them. "You see," he said, as though they should have known it all along and behaved more fittingly.

This simple little story bears repeating here because it illustrates the ambivalence of the hunter or fisherman towards his quarry and emphasizes the sense of identity with the natural world which grows out of this whole approach and which can be the ultimate reward for those who search outdoors for any natural thing.

I believe that it is important for the fisherman to reach this sense of identity and to feel it again and again in his fishing lifetime. It is important, too, for those looking after the fisheries to realize that this identification is the highest goal and that nothing that encourages it should be reduced or destroyed by the mechanics of management.

Chapter Seven The Value of the Resource

It is not very hard to see from figures alone that the recreational fisheries of Canada are a major resource. After all, some six million people, or more, take the time and trouble to go fishing in Canada each year. Collectively these anglers travelled hundreds of thousands of miles, spent hundreds of millions of dollars and supported thousands of jobs. If the anglers were properly regulated and controlled, and for the most part they will have been, they did all this at no significant cost to the resource itself. They will have scattered a few tin cans and created a little pollution of other kinds and fed a new generation or two of biting flies. They will have caught and consumed or wasted many millions of fish. Within a year most of the fish, like the biting flies, but unlike timber and minerals, will have replaced themselves and it can all be done again.

Has this activity done any good to the participants individually or to society as a whole? How can it be justified? What sense can it make in a world where some people are starving and others are facing all kinds of difficulties and dangers? I have already gone to some length to answer the first question on terms of the sport itself and its cerebral and spiritual values to the sportsman. These are reasons almost as deeply important to man as life itself. But there are substantial physical and mental values as well, and perhaps these are the best answer to my second question. No doubt the world would continue to unfold if no one had ever conceived the idea of trying to catch a fish on a hook and line, but it would certainly have been a poorer place, lesser by billions upon billions of hours and days of happiness, now and in the future.

One way or another, perhaps in spite of himself, the man who goes fishing exercises his body more than he otherwise would. Perhaps he will walk a mile or two or three to a favourite lake, perhaps he will wade a stream or row a boat. If not, he will be cranking an outboard, dropping and raising an anchor, packing stuff up and down a wharf. It is all good for him, as is the fresh air he inevitably breathes. More important still is the effect in terms of mental health. In spite of or

Churchill River, Saskatchewan

Churchill River, Saskatchewan

because of his success in making himself more comfortable, safer and more remote from the basic problems of survival, western man, and to some extent modern man everywhere, is under constant nervous strain from a multiplicity of sources, not least of them the noises from all his machines. He is under pressure to perform and produce at a rapid rate, both for himself and for others, and he does so. To be able to live under these conditions without mental breakdown, most people need some contrasting form of relaxation. People seem to recognize that going fishing provides this relaxation because they have turned to it in ever increasing numbers.

It seems to me unlikely that these lines of cause and effect, obvious though they are, can ever be measured with any great accuracy, much less reduced to useful economic terms and concepts. Yet the value to the nation and to the world in terms of renewed and increased efficiency must be enormous. The additional value, in terms of money and services saved through what is essentially preventive mental hygiene, must also be enormous.

Still another highly practical value of the resource is its power to attract people to work in industries developing the more remote parts of the country. This is well understood by industrialists, some of whom count good hunting and fishing in an area as a factor in determining the location of a plant. On a still broader plane, I have known many new and valuable immigrants to Canada who counted the prospect of good sport fishing very high among their reasons for choosing to settle in Canada rather than somewhere else.

It is clear that Canada's first and most important use of the resource must always be in the direct service of her own citizens. Although the returns may be statistically evasive the effects are clearly beneficial not only to those who actually go fishing but to the country as a whole. To express it crassly, the country needs fewer psychiatric hospitals and mental institutions because good fishing is readily available. More significantly, the sport fisheries contribute positively to the richness and quality of the lives of many Candians. In this broad sense I am satisfied that the sport has values comparable to those of the arts — music, painting, sculpture, dance, film and literature. Sport fishing is, in fact, a very real part of the culture of the country and contributes significantly to the character of its people.

Another value of the sport fishery is its power to stir Canadians not only to move around their own provinces, but to visit other provinces in search of different fishing experience. This also redistributes a certain amount of money, which I understand from economists is a good thing, but I am still inclined to place the higher value on the intangibles of Canadians getting to know each other and their own country.

Finally, there is the foreign tourist trade. A million or more anglers come to Canada from outside the country every year, mainly from the United States, but also from Europe and Japan and other countries. For the time being at least, Canada has what might be called an "exportable surplus" of good sport fishing and many of these visitors have the necessary funds to reach out for it — into the north and the far north, even the Arctic, to fly-in camps and other remote places. This traffic is valuable not only for the money it brings, but for the seasonal stimulation afforded by the visitors, for jobs provided, for resort development and for the justification of research and other government action to understand and protect the resource.

To an important extent, these foreign visitors are attracted by the quality fisheries: Atlantic salmon in Quebec, New Brunswick, Nova Scotia and Newfoundland; tuna in Nova Scotia, Newfoundland and Prince Edward Island; walleyes and brook trout in Ontario and Quebec; walleyes, northerns, grayling and big lake trout in the prairie provinces; chinook and coho salmon, steelhead and Kamloops trout in B.C.; Arctic char in Labrador and the Northwest Territories. Many of these fisheries require special care in management and few are capable of any great expansion. In some cases the entry of non-Canadians to them is already restricted by requirement of special licences.

But there will be room for non-Canadians to fish in Canada for a very long while because there are many fisheries, especially saltwater fisheries along both coasts, that are relatively little used. In the Maritime provinces one can easily imagine greatly expanded charter boat fisheries for mackerel, pollock, cod and many other fish. In B.C. ling cod and rock cod are readily available and a great deal more can be done to rehabilitate the Pacific salmons. I would like to think that sable fish *(Anoplopoma fimbria)* might one day become popular with anglers, but it seems that their habits keep them very deep in the water, and Pacific mackerel *(Scomber japonicus)* are altogether too unreliable in their appearances. Pacific pomfret *(Brama japonica)* are also considered somewhat erratic in their movements, but it is well to remember that in Florida the Atlantic pomfret is regarded as one of the most challenging of fly rod fishes.

Among freshwater fishes, the northern pike seems to be able to hold his own against very heavy fishing and on the whole is somewhat underrated by anglers. Perhaps the northern is not a spectacular fighting fish, but he is interesting and rather formidable, with a certain sudden mystery about his movements and moods, and the search for a really big one is a challenge for any keen fisherman. Northerns are not well served by the rather contemptuous common name of "jackfish"; this is an English usage usually applied to small pike of five pounds or less and to my mind does not at all fit the shadowy monsters of the northern lakes and rivers. The lake whitefish may also be somewhat underrated by anglers. In northern rivers in early spring, for instance, I have found whitefish up to four or five pounds readily taking tiny wet flies in quite swift currents. Under these circumstances they make interesting fishing and take out a lot of line when they set their broad sides across the current. In warmer weather they can be readily taken on surface flies. Big northerns can also be taken on flies — large streamer flies — but opportunities such as these are likely to be little exploited so long as more traditional fisheries hold up.

The list of potential or underexploited fisheries could be considerably expanded and it is worth remembering that refinement of fishing method can sometimes give a fishery a totally different appeal. In the same way fisheries can be "discovered" by people who come to them with a different approach; I understand that many southern Europeans who have come to Ontario and Quebec since World War II, using their traditional methods, have discovered a considerable abundance of catfish *(Ictalurus punctatus)* and carp *(Cyprinus carpio)* that provide excellent sport.

The real values of Canada's recreational fisheries are first of all in providing constructive relaxation for her own citizens, a sense of sharing directly in the country's yield, a sense of having an active part in nature and natural things. There are many obvious economic benefits in

Catfish *(Ictalurus punctatus)*
Rock cod (Copper rockfish) *(Sebastodes caurinus)*

Near Peterborough, Ontario

this and there is direct economic benefit to the country through the resource's power to attract tourists: it has been estimated that some ten per cent or more of all Canadian tourism is dependent on the recreational fisheries. But it is important to understand that this benefit is secondary, little more than incidental to the other. The recreational fisheries should not be thought of as tourist bait, nor managed for that purpose; they belong first to the people of Canada, for their own use, not to the tourism entrepreneur.

Still another benefit to the country, perhaps the most important of all, is in the need of the recreational fisheries for clean water. Much of the pleasure of fishing is in water, clean water in all its forms from mountain snow to brook and spring creek, stream and river and lake, pond and pothole and mountain tarn. The angler is very conscious of this need and of the need for properly managed watersheds to support it. He is a presence along the streams and on the lakes, a constant guardian. Sometimes he may not be as aware as he should be, sometimes he may be defeated by other destroying interests. But time and again his presence and protest have prevented abuses. His steadily growing numbers help to force cleanup of damaged waters and prevent future pollutions. Every phase of his concern, every active protest, every success, even the well-fought failures, are of long-term benefit to the country and to all Canadians.

Chapter Eight Managing the Resource

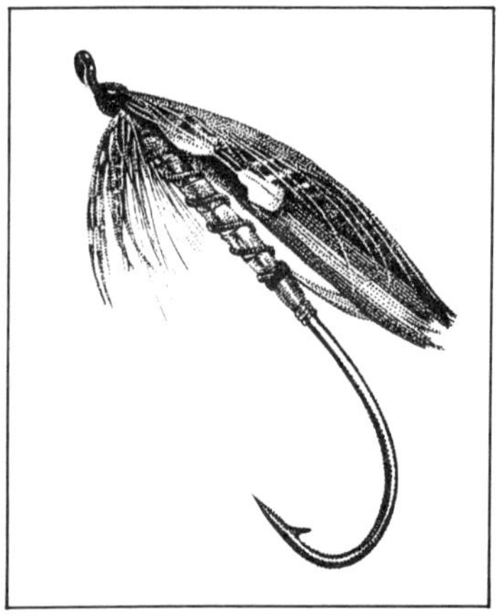

Canada's early settlers brought with them European concepts of ownership of land and water, fish and wildlife, and these were accepted with little question until Confederation and after. Riparian landowners owned the stream bottom to the centre line of the stream and so the fishing rights. Where a stream or river flowed through public lands, the Crown had similar property rights and could lease or sell the fishing rights. Superficially it seemed that the British North America Act assigned all fisheries rights to the federal government.

This simple and happy state of affairs lasted until 1882 when the Supreme Court of Canada ruled, in *R.* v. *Robertson,* that the inland, nontidal fisheries were owned by the provinces, not the Dominion, although the Dominion had exclusive power to legislate. This ruling was confirmed in 1898 by the Privy Council, which decided also that while the Dominion had a right to impose "a tax by way of license as a condition of the right to fish" the provinces also had the power to licence "in order to raise a revenue for provincial purposes."

These rulings seem to have effectively discouraged the federal authorities insofar as the freshwater fisheries were concerned, leading to a metaphorical shrug of the shoulders and a prompt withdrawal from responsibility except for the rubber stamping of legislation proposed by the provinces. If the federals felt relieved to be rid of the burden, the provinces were equally happy to seize on what seemed a promising source of revenue.

This all happened some eighty years ago and things have not quite worked out the way they seemed likely to. The recreational fisheries may produce revenue, both directly and indirectly. But they also impose important responsibilities that have largely been shirked until quite recently. The federal government has long had excellent fisheries research capacity, but until 1970 or thereabouts for the most part failed to exercise this in the interests of the recreational fisheries and similarly failed to provide significant leadership in management of the fisheries.

The provinces, perhaps with one or two exceptions, showed very little constructive interest until after World War II. In all the years before then British Columbia, for instance, had employed only one biologist and retained him for only two or three years.

The questions of fisheries ownership have gradually worked themselves out over the years, with all the provinces except Quebec and New Brunswick turning to public ownership of the resource. Nova Scotia, for instance, seems to have extinguished all private ownership by simple legislation and to have confirmed public access in "The Angler's Right to Fishing in the Province of Nova Scotia." Newfoundland recovered alienated land and water rights in the central railroad belt at a cost of 1.5 million dollars, and by the Confederation Agreement of 1949 the federal government assumed responsibility for all fisheries.

In Quebec and New Brunswick there are private holdings still standing in anachronistic elegance, and both provinces still lease many of their best salmon rivers. But Quebec is embarking on a policy directed towards public ownership by refusing to renew leases and by transferring control to local authorities. This policy is still in its experimental stages; it remains to be seen whether or not it will be successful in maintaining both the yield and quality of the province's noble salmon streams. An important part of the purpose is to reduce and eventually eliminate the ruthless and wasteful poaching of the spawning escapements that has persisted for so long. At present the poachers have a good deal of local sympathy and support and there is some hostility to leaseholders, in spite of jobs provided and money expended. The hope is that by giving local people a clear and evident share in the resource and its management these sympathies can be shifted. Poaching is rooted in the early settlers' need to live off the country and in a sense of "ownership by residence" that transcends such intruder notions as ownership by lease or purchase, so it is difficult not to feel some sympathy for it; but the present state of the Atlantic salmon stocks makes it no better than a brutal and wasteful stupidity. New Brunswick is taking the more difficult method of continuing to lease salmon waters and attempting to control poaching. The justification for this is in the financial return to the province and the important numbers of jobs provided through guiding — both of which are sanctioned by long tradition.

Both Nova Scotia and Newfoundland seem quite satisfied with the effects of public ownership of salmon rivers. Nova Scotia has relatively few nonresident anglers and feels that the very open regulations significantly curtail the illegal fishery. Newfoundland takes pride in the fact that salmon fishing is a popular sport, with some eighteen thousand or more resident anglers a year taking out individual or family licences, in spite of the fly-fishing-only regulations. An important part of Newfoundland's improvement program is a continuing policy of removing obstructions or the provision of fish passes around them, which gives the salmon a chance to spread out into more remote spawning areas and so cuts down poaching significantly.

In British Columbia the federal government has always accepted full responsibility for the salmon fisheries, controlling the harvest — both commercial and sport — and the spawning escapements, conducting research and managing protection and improvement programs. The main part of the sport fishery is carried on in tidal waters and has been open to the public without licence, except for one or two special areas, though changes are likely. The steelhead, like the Atlantic salmon, is mainly taken by anglers in fresh water and so, until recently, has been somewhat neglected by both governments. There is now a special steelhead punch card in addition to the provincial angling licence for fishing in fresh waters, and on a number of rivers, persons not resident in Canada are required to take out an additional licence. Freshwater regulations are adopted by the federal government on the recommendation of the province. Apart from these restrictions the recreational fisheries of the province are open to the public, and access, until the present time at all events, has presented few insoluble problems. There is no leasing of fisheries in the province, and I am not aware of any riparian rights except in the railroad land grant on Vancouver Island, an anomaly of doubtful status that the province would be wise to correct.

Both Ontario and Quebec have advanced provincial systems for the management of the recreational fisheries, but both welcome federal co-operation. Ontario is anxious for federal leadership in cleaning up

Steelhead from the Nimpkish River, British Columbia

Waterton Lakes National Park, Alberta

pollution in the Great Lakes. Quite clearly this involves international co-operation also, so it is certainly a federal matter. In matters involving pollution and watershed damage through provincial land and water use the jurisdiction is much less clear, not only in Ontario but in all the provinces. Section 33 of the Canada Fisheries Act spells out important federal powers to control or prevent abuses, and these have so far been respected. But no one really knows how much pressure the federal government can exert to protect the fisheries in the face of the constitutional powers of the provinces. As a result, federal powers have been exercised with caution and the effort has been to find ways of co-operation rather than compulsion. This is a weak position in the face of the ruthless traditions of logging and mining operations and the primitive waste disposal methods of too many industries and municipalities, but becomes stronger all the time as the provinces realize more fully the importance and value of the recreational and commercial fisheries.

All three prairie provinces welcome federal assistance and co-operation and have great need for intensive research on major problems, especially those where recreational and commercial fisheries overlap. These provinces, like Ontario, manage both commercial and recreational fisheries, since both are in fresh water. Manitoba feels the need for federal help with its fisheries inventory and in various research problems. Saskatchewan needs help in determining the individual life histories of its important fish species, in identifying the ability of stocks of fish to maintain their levels in balance with a fishery and in determining lake trout and walleye spawning areas. Alberta looks for further co-ordination of research and regulation, studies on species interrelationships, and examination of problems of natural reproduction in certain areas. All three provinces emphasize their need for greater flexibility in setting minor regulations and all feel that the resource does not receive a nearly adequate return of funds.

Nearly all provincial fisheries officials feel there should be a more generous provision of federal funds and, even more emphatically, that the provinces themselves are slighting the resource through meagre budgets. It is not difficult to find examples. One province, which estimates the value of its Atlantic salmon rivers at not less than 110 million dollars, is able to afford only 17 thousand dollars a year for fisheries research and improvement. Several provinces expect their conservation officers to spend a good part of their time on such duties as timber cruising, land checks and other surveys only indirectly connected with fish and wildlife. This divided responsibility grossly interferes with the performance of these officers in protection and management of the fisheries and at the same time tends to reduce the importance of such work in the eyes of the public. Such parsimony reflects a continuing failure by politicians to understand the true value and importance of the recreational fisheries.

Except for some of the Atlantic salmon streams in Quebec and New Brunswick, some early alienations, mainly in the east, and some limited use of short-term leases, Canada seems to have decided on public ownership of the recreational fisheries. In my opinion, no other decision is possible on the North American continent now or in the future. But public ownership does not mean that entry into the use of the resource can or should be free. At one time, very long ago, free entry into the resource may have made sense. Any asset has to be maintained, cared for and improved if possible, and logically this care should be paid for by its beneficiaries. Arguably all Canadians are beneficiaries of the recreational fisheries, and so it is reasonable that at least some money should come from general revenue. But the primary beneficiary, the angler, both resident and nonresident, should pay his share and pay it willingly in the form of a realistic licence fee. There is good evidence that he is willing to do so if given reasonable assurance that a fair proportion of the money will be returned to the resource; more than twenty-five years ago the hunters and anglers of British Columbia *requested* the provincial government to double their licence fee and apply seventy-five per cent of the proceeds to the resource; this proposal was accepted, the necessary legislation was enacted, but unfortunately a successor government saw fit to rescind the legislation and use the accumulated funds elsewhere — an action that has caused major conservation losses to the province. In spite of the breach of faith, I have never yet heard a complaint about the doubling of licence fees.

It is interesting, and disturbing, that Ontario, with its relatively intensive fisheries management program, has no resident licence. Even more interesting is the timidity of the federal government in failing to

Netted salmon from the Whale River, Ungava District, Quebec

Tyee salmon caught in Discovery Passage, British Columbia

require a licence for the great saltwater recreational salmon fishery in British Columbia. Some three hundred thousand anglers a year, resident and nonresident, fish for salmon in the Strait of Georgia alone. The only licence required is one for the *boats* of non-Canadians. An average season licence fee of ten dollars would yield around three million dollars which could be put to excellent use in the protection and rehabilitation of the hundreds of small streams that nurse the salmon stocks on which the fishery depends.

Most provinces do in fact have licences and licence fees and some have additional charges for special fisheries. No doubt some revenue is realized and some is returned to fisheries management — though quite clearly not enough. It would seem that this whole matter warrants thorough examination by both federal and provincial authorities with a view to establishing some reasonably consistent policy. It may well be that a single federal-provincial licence would make administrative sense and at the same time serve to emphasize the co-operative nature of the resource's management.

It has often been said that managing recreational fisheries is, in the final analysis, managing people, and there is a good deal of truth in this. Fortunately, people can only be managed to a certain extent — they must feel and understand the need for any regulations that are proposed, and generally will most willingly accept those that follow along more or less traditional lines. Any major departure calls for intelligent explanation and good publicity if it is to be successful — and that is as it should be.

Traditional regulations, familiar to fishermen everywhere, deal with size limits, bag limits, possession limits, open and closed seasons, gear regulations. Gear regulations may include fly fishing only, fly and lure fishing, or fly, lure and bait, which may or may not include live bait. Less familiar modifications, which are gradually coming into use all over the continent, include: "catch and release" or "fishing for fun" or "no kill"; single barbless hooks; daily, weekly or seasonal overall quotas for certain waters, and maximum size limits or maximum and minimum size limits. Discriminatingly used, any of these regulations may be valuable in protecting, maintaining or improving a fishery and may assist significantly in distributing or controlling fishing pressure — sometimes, but not always the same thing.

The Falls, Mécatina River, Quebec's North Shore
Matapédia River, Gaspé Peninsula, Quebec

In effect, restrictive regulations such as "flies only" or "no kill" create limited entry fisheries simply because not everyone wants to fish nothing but a fly or to release safely all the fish he catches. Such rules cannot be considered discriminatory because anyone can learn to fish a fly or to release fish safely, if he so chooses. These or similar regulations can be invaluable in maintaining a quality fishery, but they should be applied only if they fit the biological needs of the fishery. It would nearly always be wrong to suggest a fly only regulation for walleyes or northern pike or even winter steelhead; but "flies and lures only" or the prohibition of live bait are reasonable enough and could well serve important purposes. "No kill" can become an important provision in any type of fishery. Closing spawning waters during the spawning season is traditional and has a certain ethical value, even when not essential for conservation.

Quebec has already gone well beyond this point in the use of limited entry fisheries. I have already mentioned the gradual recovery of leased salmon rivers. I understand that such rivers will be opened to the public in a number of different ways — perhaps by lease to local clubs, perhaps by day licences for various pools or beats, perhaps by season licences. The present system already brings a great many public waters in the province's parks under quite strict control. The number of boats on a given lake is limited and the boats are probably government owned; on small lakes outboard motors are banned altogether and motor size may be limited on larger lakes; available accommodations or campsites will be limited and again often government owned. Boats and accommodations may be reserved in advance by phoning a central office where great wall boards proclaim the state of reservations for each area through the season. For one-day weekend reservations, Saturday fishing can be reserved only by a phone call on Thursday, Sunday fishing only by a call on Friday. There is a battery of forty telephones to take incoming calls. The demand is so great that on the first few occasions the system was used the switchboard blew out; the operators have now learned to cut the power for a few seconds after the 9 A.M. opening to prevent the pile-up of calls caused by eager anglers waiting with all but the last digit dialed.

A traditionalist might wonder what old Izaak Walton would have thought of such a system. He was wise as well as gentle and he was guided in his fishing by rules and standards of sportsmanship and in his life by ethical considerations for all men. After a moment of reflection I feel sure he would have recognized that, in this different day and age, the system is well designed to protect those qualities of peace and solitude that he himself valued so highly.

Sauble River, Georgian Bay, Ontario

Adams River sockeye salmon run, British Columbia

Chapter Nine Preservation and Enhancement: The Salmons

Few things on the planet are more fascinating than the biological problems of fisheries management. True, modern technology takes us under the water now, but we are still not at ease there, it is not our world nor likely to become so, and there is still the vastness of the oceans, of the great freshwater lakes, of even the smaller lakes. Water still hides much.

Canada has produced many great fisheries biologists and their names should be better known and better remembered than they are. I do not profess to know all the names, much less how they should be rated among their peers, but there are many names in the back of my mind and some of them come readily forward. I think of W. A. Clemens, for many years director of the Pacific Biological Station in Nanaimo, coauthor of my overworked copy of *Fishes of the Pacific Coast of Canada* (a second edition because the first got soaked). I remember R. E. Foerster, who got rid of the anadromous fish hatcheries that had wasted time, talent and money for nearly a hundred years; J. R. Dymond, who educated the Toronto Anglers and Hunters (one of Canada's earliest and strongest associations of shooters and sports fishermen) and many others of us in the process; D. S. Rawson who is still the prophet of lake productivity in British Columbia, the prairies, even Ontario's Lake Simcoe; A. G. Huntsman, who questioned everything that was not proved to the hilt; Ferris Neave, "probably the greatest pure biologist of them all," who didn't find time to complete a Ph.D. thesis until he was sixty; Paul Elson, sometimes called the father of the Atlantic salmon; A. W. H. Needler, the statesman-biologist, who has argued Canada's case so often and so well; Charlie Clay, fisheries engineer incomparable — for Canada and for the world. A long list? Not at all. Tomorrow I might offer a different list just as long: good fisheries men like my old friends Andy Pritchard and Charlie Mottley; and Gus McDonald of Saskatchewan and F.A.O. I remember Dick Miller, who died too young but left us his *Cool, Curving World.* Most certainly, on this or any other day, I would remember Vianney Legendre, that wise and courteous gentleman whose Montreal office is a jumbled repository of all fisheries knowledge — books, papers, files, documents, plans, records in seeming confusion. In the midst of it all his great leather chair is like a deer's bed in thick brush. No question is too difficult to win a gentle, soft-spoken answer, humour trembles always close to the surface and enthusiasm lights as though a youth were talking.

These men and others like them are the grand tradition of Canadian biology and fisheries management. Many younger men are working in the field today with this same spirit and the same vital curiosity. I hear them talking, listen to their excitement as I write. If the politicians give them a chance and if the fishermen give them the support they need, I believe Canada's fishing can always be among the world's greatest.

In thinking of the many problems involved in restoring and maintaining Canada's fish stocks, both recreational and commercial, it is difficult to deny pride of place to the salmons, Atlantic and Pacific. These fish must use three widely different environments: the fresh water of lakes and streams, the brackish water of the river estuaries and the salt water of the oceans. They must survive the ocean years in good numbers and complete their great migrations with almost perfect timing; and man has managed to put just about every conceivable obstacle in their way, from blocked and polluted waters to ocean nets and long lines, river nets and spears and gaffs and dynamite. When he isn't blocking the clean spawning gravels with silt he is stealing them to build roads. When he isn't tearing down the forests to ensure flash floods and low flows he is spraying them to kill every insect, harmful or beneficial, and most of the young salmon's food.

In spite of all this, Canada has a lot of fine rivers left in shape to produce Atlantic salmon — in Quebec, New Brunswick, Nova Scotia, Newfoundland and Labrador — and also the necessary base of salmon stocks. This second point is important because salmon stocks (whether Atlantic or Pacific) are peculiarly adapted to their own streams and, on larger streams, to specific parts of those streams. A native stock is always the best stock and produces the highest ocean survival rate. If a particular stock is totally wiped out it can be extremely difficult, perhaps impossible, to replace. A healthy Atlantic salmon stock is usually made up of fish that run to the river at different times, early and late, spring and fall, and at different ages, after one, two or three years at sea — and so at different sizes.

I sometimes think that all the problems of Canada's Atlantic salmon are reflected in the almost unbelievable decrease in the average size of the fish. I noticed this first of all in the records of Newfoundland rivers. In the province's fishing guide, river after river shows an average weight of between three and six pounds, which can only mean a very high proportion of the fish are grilse — salmon returning to spawn after only one year at sea. When I checked these figures against C. H. Palmer's *Salmon Rivers of Newfoundland,* published in 1928, it was obvious that this is a drastic change. Palmer reports of the Grand Codroy River: "salmon up to 40 pounds have been caught," then records a catch of 116 salmon and 14 grilse averaging 10½ pounds, taken in 1924. The present average weight for the river is about four, possibly four and a half, pounds, and the largest fish taken from 1972 to 1974 was only twenty-four pounds. River after river shows the same story. Highlands River, says Palmer, was famous for its large fish, "the pools full and not a small one in the lot"; today's average is about six pounds and the largest fish twenty pounds. Torrent River, "noted for its large fish"; today's average four and an eighth pounds, heaviest fish fifteen pounds. Repeatedly he reports "large fish up to 40 pounds" where the heaviest recorded fish for 1972 through 1974 were fifteen, fourteen, twenty-four, nine and in one instance only five pounds (Long Harbour River). This is not a new phenomenon. W. J. M. Menzies remarked on it in his 1951 report to the Atlantic Salmon Association. Nor is it limited to Newfoundland. It is almost equally true in Nova Scotia and to a lesser extent in New Brunswick and Quebec.

In excessive proportions, and in most areas, grilse are not desirable fish. Only some ten per cent live to spawn a second time, up to eighty per cent of them are males, whereas the full-grown salmon are eighty per cent females and produce the largest eggs with the best chance of survival. There is a fair probability that grilse tend to produce grilse.

The first cause of this growing imbalance is in the long history of gillnetting with too large a mesh size. In the years from 1964 to 1975 an excessive fishery off Greenland reduced the numbers of large fish and favoured the grilse, many of which do not travel that far. This fishery has now been phased down to a more reasonable level, and at least some of the Canadian net fisheries which were closed down to compensate for it, will probably be allowed to operate again. A smaller mesh regulation would permit the gill nets to catch up more grilse and at the same time allow a higher proportion of large fish to escape. Trap nets might also be required to release a certain proportion of large fish. Finally, catch-and-release regulations should be applied on some rivers, and anglers everywhere might be encouraged to release at least a proportion of the large fish they catch.

I feel fairly certain that if something of this sort were attempted over a few years the whole picture would change greatly for the better and the simple sign of it would be a sharp change in the present average weights in most streams. But this is not to suggest that good things are not now being done to help the salmon. I have already mentioned Newfoundland's removal of obstructions to open up new spawning areas in some fifty streams, with another sixty or so still to be done. Both the Humber and the Exploits rivers are expected to show a marked increase in salmon stocks within the next five years. Production on the Exploits has been further assisted in various ways, including an artificial spawning channel. Both Newfoundland and New Brunswick authorities have determined that boulder areas are more productive and able to rear more salmon parr than gravel riffles, and New Brunswick has found money in its tiny improvement budget to plant broken rock in some riffles.

A good deal of work is also being done on fish passes in Nova Scotia, New Brunswick and Prince Edward Island. In New Brunswick and Quebec part of the improvement work is undertaken by private clubs that own or lease salmon waters. Just one example of this is on the Moisie River on the North Shore near Sept-Iles. The Moisie, a big, strong river some 250 miles long, has one of the finest known salmon stocks — powerful fish averaging eighteen to twenty pounds, with many thirty- to forty-pounders, which commonly return to spawn

Salmon fishing, Nova Scotia
Discovery Passage, British Columbia

Chinook salmon *(Oncorhynchus tshawytscha)*
Brook trout *(Salvelinus fontinalis)*

three or more times. The four clubs that lease the Moisie, in addition to annual operating expenditures of over a quarter of a million dollars, have spent several hundred thousand in building fish passes and removing obstructions.

In St. Andrews, N.B., at the Huntsman Marine Laboratory, some important work in genetics is being done. The Big Salmon River, which is the next river towards the head of the Bay of Fundy from the Saint John, with a few other streams around the Bay (some of them in Nova Scotia), has an extremely unusual stock of salmon. Nearly all these fish spawn for the first time as grilse, and some return every year from then on, up to five or even eight times, to spawn again. They match this remarkable performance with an equally unusual migration pattern, travelling only a short distance from the mouth of the Bay of Fundy and keeping safely south of the Newfoundland nets and, of course, even farther from the Greenland fishery. These characteristics make them a promising stock for transplants, and a substantial number of crosses with stocks from other rivers is being attempted to see if homing instincts can be implanted without loss of the original habits.

Much effort has gone into Atlantic salmon hatcheries and smolt rearing, with varying success. There was an assessment of hatcheries in 1973, which has resulted in the phasing out of some of the less efficient plants. Smolt production will continue and increase over the next eight or ten years and should yield more favourable results with the reduction of the Greenland fishery. Smolt production is a costly business, especially when smolts must be reared over two years; it has many risks, including diseases of several kinds which can be (and some of which have been) transmitted to wild fish. The greatest value of smolt production is in the establishment, or re-establishment, of self-sustaining runs, as the successfully developing run in the upper LaHave River above Morgan Falls fishway, and in supplementing natural runs where conditions have been impaired, as at the Mactaquac Dam on the Saint John River. It is important to emphasize that hatcheries do not and cannot replace the productive yield of natural stocks. Their value is in special uses, above all in the restoration of lost or damaged runs. Smolt planting, successful though it has been, is progressively less successful in proportion to the distance of the planted stream from the donor stream.

This brief examination of Atlantic salmon problems is intended only to offer a few examples of the complexities of artificial development of the resource and its massive potential. The Atlantic provinces, including Quebec, have literally hundreds of producing streams, nearly all of which are producing below optimum levels. In 1974 the province of Newfoundland operated well over twelve thousand gill nets, many of them on or near the migration routes of salmon travelling to the other provinces; Nova Scotia had 308 gill nets and 327 trap nets, New Brunswick 93 gill nets, 188 drift nets and 404 trap nets and Quebec 479 gill nets and 348 trap nets. Many of these nets fished mainly on the stocks of nearby streams, but it is certain that there must be a good deal of interception, especially where salmon travel close along the shorelines, and in the drift net fisheries. Newfoundland estimates that only twenty per cent of its overall commercial catch is mainland salmon, though the proportion may be as high as fifty per cent off the northeast coast and twenty-five off Port aux Basques now that the drift net fishery is closed. It is difficult not to believe that really close examination of these fisheries would suggest a system of controls that could be of immense value to net fishermen and anglers alike. Without such control full restoration of the runs will always be an uphill task and may be impossible in the case of some stocks.

On the Pacific Coast salmon problems are quite different, not so much because of the life histories of the fish but because the salmon run in such massive numbers and the commercial values are so high. Research assumes different proportions, management is on a different scale in controlling the fisheries and spawning escapements, and enhancement efforts are likely to be in the same scale. Of the five Pacific salmon species, the sockeye, the pink and the chum are the main support of the net fisheries. The pink salmon goes to sea immediately upon hatching and returns after two years, fully mature. There is no freshwater life history as such, though estuarial conditions may be critical. The sockeye, except in more northerly parts, is a four-year fish and must spend one year in a freshwater lake. The chum salmon is

Blueback (young coho salmon) *(Oncorhynchus kisutch)*
Coho salmon *(Oncorhynchus kisutch)*

Discovery Passage, British Columbia

somewhere between the other two species, maturing at three, four or five years but spending very little time in fresh water, though again estuary conditions may be critical to satisfactory survival rates. All three of these species can be immensely assisted by artificial spawning channels, which provide ideal gravel and flow conditions and yield very high survival rates from egg to fry. Sockeye stocks cannot be increased beyond the lake rearing capacity of the particular section of the watershed they home to, and improvement of pink and chum stocks may, as I have suggested, be limited by estuary and early sea conditions. But even so, the pattern of enhancement is perfectly clear and logical and will certainly succeed.

Sockeye, pink and chum stocks are chiefly commercial, though pink and perhaps sockeye salmon are likely to enter increasingly into the sport fishery. The chinook and coho salmons are classed as game fish, though both support extremely important commercial fisheries as well. Both species, like the Atlantic salmon, are dependent upon a period of river life before going to sea, the chinooks from ninety days to a year or more, the cohos for one year. A third anadromous game fish, the steelhead, has a life history almost identical to the Atlantic salmon; most B.C. steelheads spend either two or three years in the rivers before going to sea, though a few spend as much as four years.

As soon as the matter of river life comes into the question, simple spawning channels, so effective with other species, will not necessarily be adequate. As there must be lake rearing capacity for the sockeye salmon, so there must be stream rearing capacity for chinooks, cohos and steelheads. Pacific Coast streams are not extremely fertile and whatever rearing capacity they have has all too often been damaged by logging operations and by industrial or other commercial encroachments upon the estuaries.

The original centre of abundance of the big sea-running rainbows is probably the Columbia River, between the states of Oregon and Washington. Abundance falls off quite noticeably both northward and southward of the Columbia, and the British Columbia streams have never been nearly so productive as those of Washington State, probably because of the colder water in the Canadian streams. In California a high proportion of steelhead go to sea after one freshwater year; through Oregon and Washington two-year-old smolts increasingly predominate and in southern British Columbia yearling smolts are rare, while two-year-olds make up forty per cent and three-year-olds fifty-five per cent of the runs. These late migrating smolts are also likely to spend a longer time at sea and so return as larger fish.

The steelhead is not a highly valued commercial fish. Incidental interceptions by salmon nets are a serious problem, especially on large watersheds like the Fraser and the Skeena. Angling pressures are very high on all but the most remote streams and spawning escapements are generally inadequate. Artificial hatching and rearing, apart from such disadvantages as genetic pollution of wild stocks and the very real hazard of disease, is made difficult and costly by the low water temperatures. There is an excellent case for catch-and-release fisheries on many steelhead streams.

Pacific Coast salmon stocks, including the two game species and probably also the steelhead, in 1975 were producing at rather less than half of their historical potential. Much of the falloff is due to hydroelectric and other dams, to bad logging practices, estuary damage and pollutions. Enhancement programs designed to double present catches now are proving successful. Sound use of spawning channels alone, as I have already suggested, can, and on some rivers has, certainly doubled stocks of sockeye, pink and chum salmon. Successful improvement of coho and chinook stocks, and with them the steelhead stocks, will depend mainly on imaginative and aggressive stream improvement and protection work, perhaps with limited use of hatcheries and spawning and rearing channels. Coho salmon stocks especially are dependent on stream improvement work; the opportunity is a great one. Around the Strait of Georgia alone there are literally hundreds of small, often very small, coho spawning creeks and streams. Most have been damaged by logging and other abuses but most now have well-restored ground cover. The possibilities for improvement are immense, and the effort already under way should advance the technology of stream improvement throughout the world.

If I sound optimistic about the possibilities of salmon restoration and development on both Atlantic and Pacific coasts it is because these fish, within their rather specific requirements of clean waters, good flows, clean gravels and productive estuaries, have a resilience and adaptability that will reward any serious attention. It is urgently important to protect natural stocks, to provide accessible habitat of high quality, modelled as closely as possible on productive natural habitat, and to ensure adequate spawning escapements. Perhaps I should add that it is also essential to protect the spawners once they have escaped upstream. This is vitally important on both coasts, but especially so on the Atlantic side where illegal upstream fishing can slow down or defeat restoration efforts. After that, the fish and the ocean will do the rest.

It is often said that in Canada's recreational fisheries as a whole the salmons are not all that important. Although that is perfectly true, it is also true that the salmons hold a special place in the angler's world of fishes. "The Salmon," wrote Walton, "is accounted the King of fresh-water fish" and most anglers would still agree, though many Pacific Coast fishermen would add: "and of saltwater fish also." This special appeal lies not only in the salmon's qualities as a game fish, but also in his romantic life story and his great beauty of both form and colour. If I have given a lot of space to his affairs in this short book I am merely reflecting one more worthy tradition of the sport.

Chinook salmon from the Atnarko River, British Columbia

Arctic grayling, Northwest Territories

Chapter Ten Preservation and Enhancement: Other Species

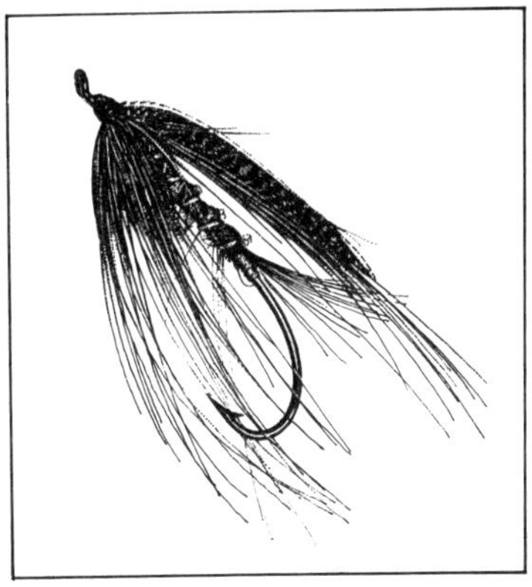

In addition to the salmons and the steelhead, Canada has two other sea-running trout, the cutthroat of the Pacific slope and the imported brown trout *(S. trutta)* now established on both coasts, and at least three sea-running chars: the speckled trout, the Dolly Varden and the Arctic char. All five are important game fish and none of them is as well understood as it should be for sound management, though a great deal of important work is being done with Arctic char, the colourful fish of the far north, at once a lure to anglers, an important food fish, a source of jobs and a valuable commercial fish.

Both the coastal cutthroat and the sea-run brown trout have somewhat limited seaward migrations, probably extending little beyond the influence of the home stream. The cutthroat, like the coho salmon, is a frequenter of very small creeks as well as larger streams, and will certainly benefit from the stream enhancement work now being done and even more so from estuary protection. Coho salmon and cutthroats seem able to thrive in the same small streams, even though there is heavy competition for food in the pre-migrant stage and each species preys on the other.

Brown trout seem to establish sea-running populations wherever they are planted in streams open to the sea. They have done so in Newfoundland where they were first introduced more than ninety years ago and in the Cowichan River on Vancouver Island where they have been established for over forty years. They are quite important fish in Newfoundland and the Maritime provinces; Nova Scotia claims an average weight of five to seven pounds for sea-run browns, with maximum weights of ten to fifteen pounds.

In spite of the popularity of salmon fishing the speckled trout is the main interest of Newfoundland anglers, as it is in the other Atlantic provinces, including Quebec. On the whole, except perhaps in Quebec, not nearly enough attention has been paid to the speckled trout or to the other important inland species, including the ouananiche or landlocked Atlantic salmon, the lake trout and the Arctic char. One Newfoundlander told me that the inland fisheries "need a good shot in the arm." He meant that they should be given the kind of research and all-round attention that has been given to the salmon and the saltwater commercial species.

Speckled trout are the most important game fish of Prince Edward Island, where they are found in good numbers in every pond and stream. There is also excellent estuary fishing for sea-run speckleds running from three to five pounds, though I understand careless handling of fertilizers and insecticides has damaged this fishery. Of all the provinces, Quebec sets most official store by the speckled trout. It may well be that Quebec is the centre of abundance of the species, and there can be no doubt about the province's determination to regard the speckled trout as a sacred trust, its natural stocks a resource to be protected in every possible way. In talking of this with Quebec authorities over the years, one feels a sense of vision and emotional involvement felt elsewhere, though less intensely, for the Atlantic salmon. In the central zone of the province, north of the forty-ninth parallel, stocking is allowed only with the progeny of local native brood stocks. In the northern zone there will be no stocking whatsoever except with specific government approval.

Saltwater fishing in the Atlantic provinces, though important, is somewhat overshadowed by the freshwater fisheries and perhaps underexploited. The bluefin tuna is a world conservation problem and perhaps little influenced by the relatively small catch in Canadian waters, but I believe some thought has been given to the possibility of releasing rod-caught fish. The main obstacle seems to be the commercial value of the fish, which can be much more than the price of the boat charter.

Some fisheries scientists believe that stocks of pollock have been damaged by pollution in the Bay of Fundy and along the North Shore, but most stocks could withstand more fishing, and these fish, which run up to thirty-five pounds, can provide excellent sport. Pollution from the St. Lawrence has also affected stocks of tomcod *(Microgadus tomcod)* and striped bass. Both species are highly sensitive to pollution, and I understand the spawning populations that used to move into the St. Lawrence are now seriously depleted. On the whole the striped bass has not attained the same popularity in Canada as a sport fish that it has in the United States, but its time will certainly come, as will that of the pollock; it is important to protect the stocks now from any further losses.

Some of Canada's most popular freshwater game fish, like the yellow perch, walleye and northern pike, are thought to be able to "look after themselves." This has proved a dangerous assumption in many prolific fisheries and no one should be deceived by it. There is already a feeling in some areas that walleyes may be suffering from overfishing; the heavy commercial catches that were the rule from Lake Erie and the prairie provinces have fallen off, and most fisheries are now controlled by quotas and mesh size regulations. It seems reasonable that anglers should do their part by abandoning the practice of fishing the spawning run. Another popular species, the smallmouth bass, has been the subject of concern for over a hundred years, It is now controlled and regulated as a game fish and yields a catch of between seven and eight million fish a year to Ontario anglers. Since its distribution is in the southerly, more populated parts of the country, it is considerably threatened by pollution.

Hatcheries, transplants and the introduction of exotic species are favourite toys of angler theorists and to a lesser extent of fishery managers. Where potentially productive waters are empty of fish, as in the case of many of the Kamloops lakes of British Columbia, the ponds and potholes of the prairie provinces and some of the lakes of Ontario and Quebec, transplants from nearby natural stocks make obvious sense and usually yield spectacular results. Similarly the introduction of such exotics as brown and rainbow trout to the largely fishless streams of New Zealand, Chile, Argentina and other countries of the southern hemisphere has proved an immense success. But I sometimes wonder if, apart from these rather special cases, the whole history of transplants and introductions and much of the history of hatcheries has not added up to loss rather than gain to the fisheries of Canada.

Artificial production of fish will always have a place for research purposes, for the stocking of heavily fished waters near large population centres, such as the Dartmouth Lakes near Halifax, for the establishment of self-sustaining populations in lakes without fish, for support of existing populations that lack adequate spawning areas, or for mitigation, as at Mactaquac Dam on the Saint John. But hatcheries are a management tool to be used with caution. Planting hatchery fish in waters capable of sustaining adequate wild populations must always be wrong. Progeny of hatchery brood stocks that have been specialized through several generations should never be used in any attempt to develop a self-sustaining stock, though they may be satisfactory for pond culture. The idea of transplants will always be tempting: as an angler theorist myself I cannot help feeling that Canada has not looked well into the potential of its landlocked Atlantic salmon. I know it has been a problem to find a source of eggs, but a truly diligent search might find one, probably in Newfoundland. The ouananiche is a noble fish and surely is a better prospect for planting east of the Rockies than the Pacific salmons.

Many of Canada's recreational fisheries are north of 55° latitude. These fisheries raise many difficulties; the extent and remoteness of the area is a challenge in itself. Conflict, or seeming conflict, between commercial and recreational fisheries needs constant watching and control, and the slow growth rate of the northern fish makes all planning more complicated. This last may well be the central problem, reflecting back on all the others, if it does not create them. Since growth is pretty much a simple function of climate, there is little to be done except understand it and work with it — and that is a very great deal.

In southern Canada the salmons, Atlantic and Pacific, and the trouts reach maturity at between two and five years and few live beyond seven or eight years. Northern pike in Georgian Bay probably mature at three to five years and may reach forty inches in length at six to ten years. In the north they will not mature before six or seven years; a forty-inch pike in Waskesiu Lake (central Saskatchewan) will be fifteen years old, in Athabasca Lake perhaps twenty years. In Great Bear Lake such a fish would probably be at least twenty-four years old, perhaps even forty.

Pangnirtung Fiord, Northwest Territories

Northern pike *(Esox lucius)*
Lake trout *(Salvelinus namaycush)*

Lake trout mature at six or seven years in the south and at twice that age in Great Bear Lake. They will grow to thirty inches at fifteen years in Algonquin Park, at thirteen years in Lac la Ronge, fourteen years at Great Slave Lake and twenty-four or more years at Great Bear. Fish beyond that age in Great Bear may be anywhere from twenty-seven to thirty-eight inches in length and from eight or nine pounds to forty pounds in weight. After a certain age, the fish grow very little, if at all.

Clearly, the stock of really large fish (northern pike over twenty pounds, lake trout over thirty pounds) in most northern lakes is made up of very old fish. In the early stages of a fishery there may be substantial numbers of large fish, built up very slowly over the years. As the fishery increases, the number of big fish will be quite quickly reduced, and since most of the fish just under the maximum size will also be caught, the chances of the big fish being replaced are very small.

The northern fisheries attract fishermen who are looking for trophy fish and are prepared to pay good money to find them. These fishermen are catered to by fly-in lodges and camps and are likely to come back year after year so long as the big fish are around. But when the size of the potential trophy fish drops in one area — say from thirty pounds to twenty pounds — many of these anglers will move on to lakes that have been more recently opened. Since these are likely to be farther north, with still slower growth rates, it will not be long before they too begin to show a drop in size. The sport fishermen blame the commercial men for this drop and the commercial men blame the sportsmen. The simple truth is that both are responsible, with the main onus probably on the sportsman because under these conditions his fishery is quite efficient and he is deliberately seeking out the big fish.

The fishing camps and lodges are important throughout the north. They employ a good number of local people both in general service around the establishments themselves and as fishing guides. It is obviously undesirable from every point of view, and I think unnecessary, that the quality fishing which led to a lodge's establishment in the first place should be allowed to pass on. What is needed is a system to protect the fish that will shortly grow to trophy size and permit the lodges to provide reasonable access to big fish. This could be achieved by closely calculated maximum and minimum size limits. For instance, in a lake that produces trophy lake trout of thirty pounds and over, all fish between fifteen and thirty pounds might be released. This would remove some of the competition by reducing the numbers of small fish and leaving a substantial population of big fish to grow up to trophy size. Limits of this type would call for rather exact examination of growth rates and age levels if they are to be truly effective and would probably vary from lake to lake. Individual anglers could carry the idea a step further by releasing any fish smaller than their own best trophy fish. If anglers and guides are properly instructed, weighing the fish before release would present no difficulty.

In most parts of Canada the food fisheries of the aboriginal people have an absolute priority. On the Fraser River, for instance, native food fisheries are carefully calculated into the upstream escapements through the commercial fisheries. In northern Saskatchewan the fisheries, both commercial and sport, are run by the separate Department of Northern Saskatchewan and major emphasis is on the benefit of the thirty thousand people resident in the area, about two thirds of whom are of native ancestry. In Manitoba there is no restriction on the native food fishery, but no positive priority. Domestic use has absolute priority throughout the Northwest Territories; in fact there is no external authority over this use, though the Inuit are co-operative with management requirements if these are properly explained.

On the whole, sport and commercial fisheries need not be bitterly competitive in the north or anywhere else, and generally they are not. In the British Columbia salmon fishery, for instance, the sport fishery in terms of fish used is a mere by-product of commercial fisheries management, except in the Strait of Georgia.

On the Atlantic side the relationship between angling and commercial fisheries is well shown in the records of the Moisie River District as given in Edward Weeks's fine book, *The Moisie Salmon Club*. In the years from 1902 to 1970 the nets took 347,966 salmon, while rods took 47,863. The best year for the nets was 1908 when they took 14,013 fish to the anglers' 507. The biggest year for anglers was 1926, when they took 2,026 fish against 10,422 for the nets. In 1970, with the

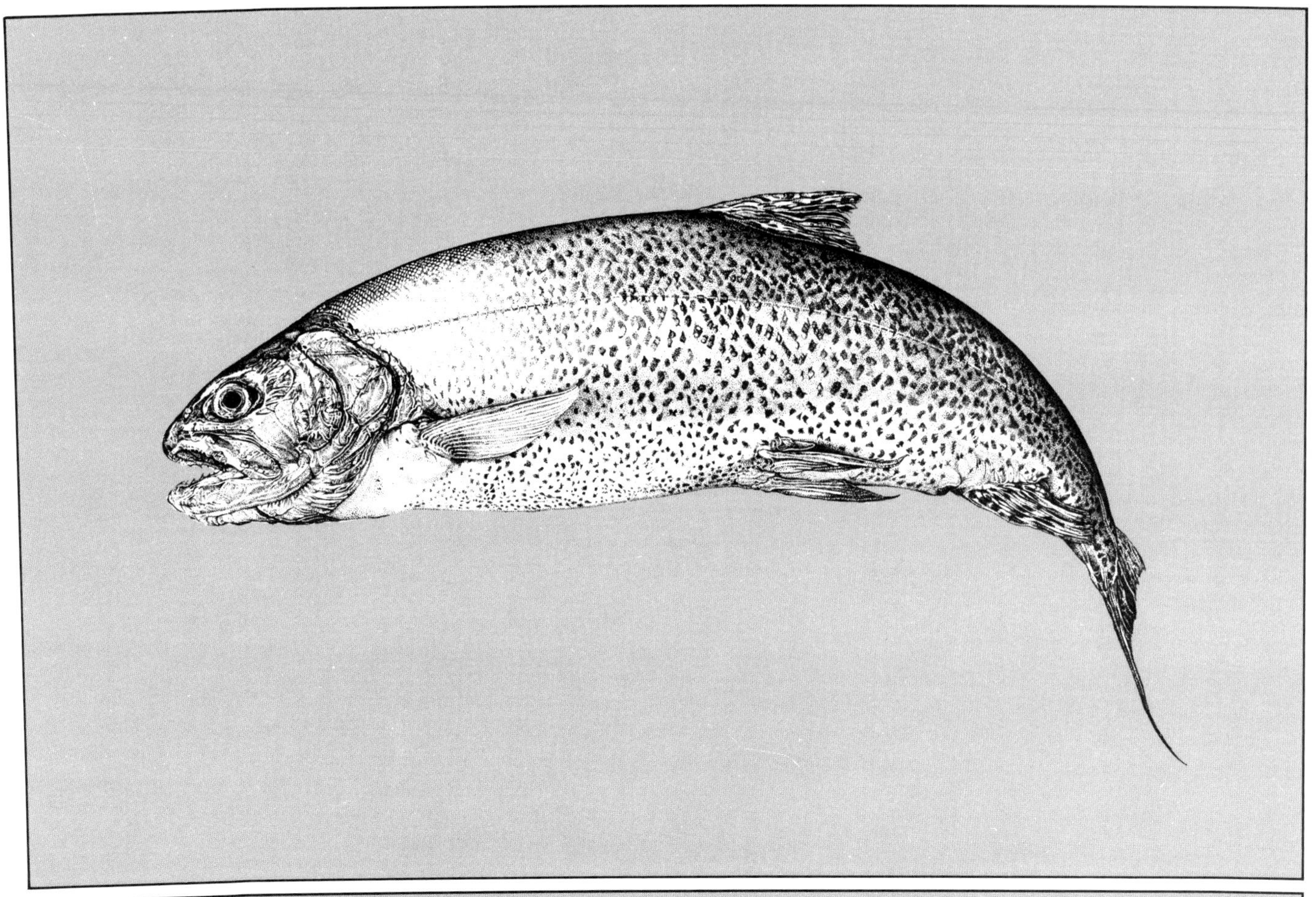

Cutthroat trout *(Salmo clarki clarki)*
Arctic grayling *(Thymallus arcticus)*

Island Lake River, Manitoba
Churchill River, Saskatchewan

Greenland fishery in full operation, the net catch was 1,920 fish, the rod catch 335.

In the prairie provinces and in the north things are more difficult, again because of the slow growth rates of the fish. Commercial fisheries are managed by quotas, by allowable percentages of sport fish in the catch and by seasons, with considerable concern for the sport fisheries. It has now become clear that the value of the freshwater recreational fisheries greatly exceeds that of the commercial fisheries. To a considerable extent the two fisheries are interchangeable — commercial fishermen often turn to guiding in the tourist season and greatly enjoy the change. In Manitoba native-run lodges are said to be excellent, and there is good reason to believe that northern Indian people in all the provinces will turn increasingly to this type of enterprise. I am told by northern people that guiding is a well-liked and well-respected occupation among native people and this view certainly agrees with my own experience.

There is a fairly general consensus among fisheries managers across Canada that both sport and commercial fisheries are needed to keep the resource in good balance, but this will work well only if there is intelligent and fairly intensive management of both. The proposition is well examined in Dr. Ward Falkner's paper, "Fisheries Zoning." Dr. Falkner suggests careful assessment of all the relevant factors, including existing fisheries, the characteristics of the fish stocks and economic returns. In the light of this information the provincial waters might be categorized as commercial, multiple-use or one or other of three types of sport fishing zones. An example of this sort of management already exists in the Northwest Territories, where the east arm of Great Slave Lake has become essentially a sport fish reserve. Elsewhere in that enormous lake the lake trout have become quite scarce, even though the west arm of the lake is generally more productive, with a particularly good yield of whitefish.

In some ways the most interesting challenge of all is management of the Arctic char. The Arctic char is of major importance to the Inuit in the Northwest Territories as a food fish, and of considerable importance there and along the Labrador coast as both a commercial and a

Pangnirtung Fiord, Northwest Territories

sport fish. It is easily overfished because of the slow growth rates, but it is also a resilient fish which can recover quite quickly if given adequate protection. There are both sea-running and freshwater resident stocks of Arctic char, but the sea-run stocks are more important both to the native people and to the anglers because of their greater size. Growth rates vary considerably according to latitude, but the young fish in the Arctic spend their first five or six years in the rivers and go to sea for the first time at six or eight inches in length. They may spawn for the first time at eight or nine years, and the most important fish in the commercial fishery are those from eight to fifteen years of age.

Arctic char command a high price as a luxury food, both frozen and canned. There is a small Inuit processing plant at Rankin Inlet on the west side of Hudson Bay and small commercial and food fisheries at other inlets and on some of the Arctic islands. It appears that most, if not all, sea-run stocks in the Arctic spend the winter in fresh water, whether or not they are ready to spawn, but only some of the rivers they enter are spawning rivers. The spawning rivers are critical and must be fished with great care, on a sustained yield basis. Management is by quotas with the average size of the fish as the guide — a drop in average size signals that stocks are being depleted and quotas must be reduced or the river rested altogether for a few years. The commercial fishery in Rankin Arm, for instance, can take forty-five thousand pounds a year from one of its streams and maintain an average size of eight pounds, but a catch of eighty-five thousand pounds would reduce this average noticeably.

Not all Arctic char rivers are angling rivers. In some it appears that the fish cannot be persuaded to take fly, lure or bait of any kind. It is hard to tell whether or not this is a problem that may one day be solved by some ingenious fisherman, but my own limited experience of Arctic char suggests that something of the sort may happen, and also that in some areas a saltwater fishery near the river mouths may be a possibility. But the real problem is not so much one of catching the fish as of controlling the catch. The Tree River in Coronation Gulf provides a good example of the delicacy of management needed. The fish take well there and in spite of the remoteness of the place the initial fishery was far too heavy and the size of the fish dropped off. A five-hundred-fish angling quota has restored the normal eight-pound average; the quota may shortly be increased to seven hundred fish, but the effect will have to be closely watched. The Tougat, an overwintering river in Pond Inlet, has an experimental quota of five thousand pounds for the commercial fishery, with a daily limit of four fish and possession limit of seven for anglers. This is considered conservative, but again the effect must be closely watched.

The far north — the Territories and all the Arctic islands — makes an immense country and there should be room in it all for plenty of fine fishing without serious threat or damage to the natural stocks of Arctic char, just as there is room along the Labrador coast. But I find it hard to understand why anglers, travelling so far to know the rich northern experience and finding this remarkable creature that has learned to live with the hard climate, should want to kill fish. Surely the Arctic char rivers are a place for catch and release, the single barbless hook, with provision for the taking of an occasional trophy fish and perhaps a fish or two for the angler to carry home — two eight-pound Arctic char should satisfy anyone. Since Arctic char make little if any growth after they are twenty-three or twenty-four years old (and have been known to live to forty and even sixty years), no doubt a proportion of the older and larger fish should be taken, and some of the smaller fish should also be available, since the Arctic char has few predators, and survival rates, as indicated by tag recoveries, are very high. Perhaps this is another place where maximum and minimum size limits would make sense.

Management questions of this sort are very much the concern of the angler. If he is at all serious about his sport he should recognize the problems, and his concern should always be to err on the side of generosity — to the fish and to the resource as a whole.

River fishing, Yukon Territory

St. Andrews, New Brunswick

Chapter Eleven A Hope for the Future

I have limited this discussion of the recreational fisheries resource to angling, but there are other important aspects. For many people an important part of a holiday on either coast is in picking, gathering, digging or catching shellfish — oysters, clams, mussels, abalones, scallops, crabs. Wandering the beaches is a pleasure in itself, made sharper by the special objective, while the seafood that results has a quality and freshness almost impossible to equal in any other way. Enthusiasm for this activity is already great enough to call for bag limits and special seasons. Shellfish are under constant threat from pollutions, both domestic and industrial, and the need for watchful protection becomes greater all the time.

Still another recreational use of the fisheries resource is simply watching fish. The popularity of aquariums everywhere is good evidence of the interest people have in seeing live fish in the water, and seeing them live and free in their natural surroundings has even greater appeal. Spawning salmon offer some of the best opportunities for observation, and because of the sheer numbers of fish in the runs, the Pacific salmons make the most impressive shows. When the chum and coho salmon are running to the tiny Goldstream, just outside Victoria, it is not unusual for four thousand cars to bring visitors to Goldstream Park on a single day. The people of Victoria, whose water supply depends partly on the Goldstream, set such store by these runs that in a year of low water they made special sacrifices to ensure that the salmon had enough. Thousands of people come each spring to watch huge rainbow trout from B.C.'s Kootenay Lake spawn in the Lardeau River below Trout Lake. At Meadow Creek spawning channel and Redfish Creek on the west arm of the lake thousands also watch the spectacle of kokanee salmon spawning in early fall. In 1974, 36,000 people visited the International Pacific Salmon Fisheries Commission's sockeye salmon spawning channel at Weaver Creek, B.C., and more than 125,000 walked the banks of the Adams River to see a million scarlet fish milling and spawning over the bright gravel in the river's clean water.

These examples represent unusually favourable conditions, but there are plenty of other places across the country where fish can be watched and it would be a simple matter to develop still others — a worthwhile possibility for conservation reasons alone, because nothing enlists the support and understanding of the public more surely than seeing the fish as they really are.

Fish watching in the fish's own element has become another popular practice as the arts of scuba diving and snorkelling become better known. A wet suit is usually needed for cold Canadian waters, but that in itself is something of an advantage, since it gives the swimmer greater buoyancy and permits him to rest and relax in the water while watching. I believe this interest will continue to grow and again there will be need for protection, especially in salt water. It is time already for Canada to be looking for special areas that would make valuable underwater parks and classifying at least some for permanent protection. In spite of its popularity, spear fishing is not a wise use of the resource; it destroys the essential peace and the absence of fear that exists between man and animals under water and, apart from essential research, contributes nothing at all.

Any opportunity to watch fish is an educational experience, no matter how casually it is taken. At the Adams River and Weaver Creek and Goldstream there are displays that reinforce the experience by explaining what is going on in the water. Anglers need a little more education than this, even when they have been carefully raised to the sport by angler parents, and on the whole it is best found in the anglers' organizations: fishing clubs, rod and gun clubs, the wildlife federations and outdoor groups of various kinds. These are good learning places and at the same time good teaching places. All outdoor groups have developed enormously in knowledge and sophistication over the past twenty years — they have had to in order to make any show at all in attempting to defend the resources they depend on. Not infrequently, outdoor groups come to public hearings better prepared and with better briefs than the developers.

An angler does himself a favour by joining a group, not only for the fellowship and the chance to learn something but also because he adds his own strengths to that of the group and has a ready opportunity to express his own views in the way that will do most good. Organized outdoor people can contribute a lot to management, just as they can learn from administrators and biologists, and they are increasingly listened to, though governments and individual politicians would do well to pay more attention than they sometimes do.

Increasing environmental awareness has had its very noticeable effects in schools everywhere and many schools have access to environmental centres of one sort or another. But I should like to see every school in the country adopt some natural area as its own special project, continuing year after year in perpetuity. Preferably it should be a creek or some section of a creek or small stream, but it could almost equally well be a section of river bank or lake shore or sea shore, a pond, pothole or marsh. The effort should be to know the place intimately, in fullest detail, to note every change, from season to season and year to year, to learn every possible circumstance about the creatures that frequent it, from algae and insects to fish and mammals, to keep it clean and free from litter, to protect its banks if necessary from flood or other erosion, to protect it from encroachments so far as possible. I see a program of this sort as having a double function: that of developing in children the fullest awareness of a given area, with a sense of possession and involvement, and at the same time providing a substantial measure of care and protection for the place itself. A program of this sort would have its greatest effect if adopted as part of the regular school curriculum of a whole province, as has occurred in British Columbia, where the Salmonid Enhancement Program's educational packages are in active use.

I am not sure I have properly emphasized the "close-in" fisheries — those near large population centres which are intensively used by all sections of the public. At times Winnipeg's Red River entertains as many as four thousand anglers to the mile. The sea wall around Vancouver's Stanley Park is constantly used by anglers, as are several lakes near the city. But its north shore streams that were once so popular have largely been ruined by development, including gravel removal and gravel washing, though perhaps they are not beyond recovery. Ontario puts out a small folder on places to fish close to Toronto and another listing a dozen provincial fishing areas, which are described as "part of the continuing program to give urban residents a close-to-home fishing opportunity."

Ice fishing is as popular in Quebec as in the other provinces. Yellow perch and tommycod are the fish usually taken in this fishery, but for some reason the traditional fishing methods do not take whitefish well, in spite of their abundance. With the resourcefulness that is typical of the province's management, a very senior biologist was sent to Lake Simcoe in Ontario to learn the finer points of catching whitefish under the ice and bring word back to Quebec anglers. It seems abundantly clear that all the provinces are well aware of the need to provide these opportunities and will continue efforts to provide them.

Something should be written of the management failures in the history of the resource. For much too long, as with other resources, all was failure and neglect, because of the quaint assumption that the resources of this continent were inexhaustible. In the case of the fisheries resource there was a further weakness in the uncertainty of responsibility between federal and provincial authorities, and in the case of the recreational fisheries the absurd concept that these were a minor luxury, the concern of few, to be written off in the face of "more important things," usually exalted with the name of the bitch-goddess, Progress. In Newfoundland a provincial fisheries man told me bitterly that the resource "has been sacrificed to the pulp companies" — an overstatement, perhaps, but with a powerful ring of truth that applies to other provinces as well.

These abuses and so many others like them — the needless waste of lands, forests and fisheries because of hydro projects, the uncontrolled and unnecessary pollutions by mining and industry, the ruthless logging operations and destructive spraying of the forests — are, one hopes, errors of the past, not to be repeated. Not all are irreparable, though repair may be slow, painful and costly. Ridiculous as it seems in a society considered advanced and at least partially civilized, it has always been necessary to meet industrial abuses, whether proposed or active, with values measured in dollars. Fortunately, fisheries economists are constantly refining ever more sophisticated ways of providing and supporting figures of this sort and forcing developers

Ice fishing, Lake Simcoe, Ontario
Upper Fraser River watershed, British Columbia

Salmon in the home pool, Jupiter River, Anticosti Island, Quebec

back to their drawing boards, but dollars remain a very questionable measure of real value.

These failures were not really the ones I had in mind. It is well to recall that there have been direct management failures of the fisheries themselves. The Atlantic salmon failure is deeply rooted in the historical context of excessive netting rights and weakness in applying controls; recovery will be hindered by the persistence of the old rights and by resistance to regulation, but there is little doubt that it can be achieved and maintained. Pacific salmon failures have been heavily associated with land abuses of all kinds — with dams and badly planned road and railroad construction. But there was much overfishing of various stocks and species and, for a very long time, failure to recognize the full importance of freshwater habitat. Even today there seems to be a continuing reluctance to become deeply involved in habitat enhancement, though there have been immense advances in understanding what the fish need and how to provide it. While there will always be an uphill fight to control careless developers, most salmon stocks can be restored to double their present yield, and the promised federal-provincial multimillion-dollar enhancement plan should achieve this.

In the prairie provinces management has failed to maintain the commercial yield in lakes such as Peter Pond and Reindeer and even in some parts of Athabasca, where there is a suspicion that forty or fifty years of fishing may have selected out superior genetic strains. Lac la Ronge, on the other hand, has been fished commercially almost since the start of the century. Until road access was completed in the late 1940s there was no noticeable change in the fish populations, but the size of the fish began to decline almost immediately after, so in this case it is angling that has been inadequately controlled. Such lakes as Nipawin and Wollaston will bring similar fisheries problems unless accurate angling controls are promptly established. Arctic char fisheries have been fished too hard in places such as Frobisher Bay, and commercial fishermen say that in many lakes it is hard to find "jumbo" whitefish — that is, whitefish over four pounds — any more.

Most errors and failures in management have come through pressing too hard at the tolerance of the resource: trying to develop a fishery too fast and harvest too much. Typical examples are the crash of the B.C. herring populations, the overfishing of steelheads in most B.C. streams, the size reduction of trophy lake trout and northern pike in so many lakes, and the unexplained fluctuations of walleye populations in some places. Clearly there is already a much more cautious and conservative attitude towards both commercial and sport fish harvesting, but these past failures are worth noting because there is still room for others like them to occur. Only comprehensive inventories backed by sound research and constant vigilance can hope to prevent them, and this success will nearly always call for a high degree of federal-provincial co-operation.

Failures can usually be corrected — in time and with difficulty — by improved management, which can often involve total closure of a fishery for a substantial period. Damage resulting from bad land management and bad land use is much more difficult to repair and in some instances may be permanent. In certain types of land the ravages of strip mining can only be repaired by the natural effects of thousands of years. The destruction of productive water by high dams is permanent. We do not know, even yet, how difficult it may be to replace a local race or stock of anadromous fish wiped out by pollution or reckless logging or some obstruction incidental to domestic or commercial activity; we only know that replacement can be slow and difficult, often resulting in many years and many million dollars of lost production. We do not know how much damage we have done to the productivity of our estuaries and marshlands, both fresh water and salt, nor have we yet learned what we need to know about the restoration of estuarial values.

What we need and must somehow find in this last part of the twentieth century is a land and water ethic — perhaps better, an ethic of land, air and water. It is perfectly possible to have settlement, industrial development and reasonable exploitation of primary resources without condemning our children to generations of poverty and deprivation because we have ruined the land that should support them. What it will take is an attitude of mind, an understanding that blacktop destroys, dirty water destroys, bad air destroys, and a solid determination to control or prevent abuses even if it seems to cost something in immediate material effects. I believe we *have* a changing attitude, a developing ethic that can achieve this, just as we already have all the essential technology to achieve it. All ideas have their time and the time for this one is well in sight.

It is exciting to think of the infinity of experiences the Canadian recreational fisheries can offer. I think of a man standing hip-deep in water at some creek mouth on the Strait of Georgia, a high tide flooding in and an occasional coho salmon swirling within fly-rod reach on the lavender-grey water. Few places are more deeply peaceful, even now when the sound of seaborne and airborne motors is always on the ear. I remember, too, at dawn or dusk, in estuary or river pool, the formidable power of a great chinook salmon that has taken solid hold of a lure and surges strength against the left hand on the rod; a cast of fortune seized by fortune, the whole outcome still uncertain.

I am not likely to forget the dark water of the winter steelhead river breaking heavily against my waders, plunging past me into the holt that my fly is searching or the quiet rise of an explosive summer fish to a floating fly. It is important also to remember the search for a great pike in the mysterious waters of Cree or Wollaston or Reindeer lakes, wide, wild northern places where anything can happen and the peace has all the uncertainty of true wilderness. I recall the happiness of the Fond du Lac River at the head of Athabasca Lake, with its big rough rocks that gave sure hold to the searching foot on a perfect August day; the Arctic grayling, that flower of fishes, so beautiful that the eye can scarcely accept it, taking freely, arcing out to the dance of the surface fly on the swift and broken water. On that day the whole place matched him in scale and gentleness, although few human travellers had passed that way since the last of the fur-traders. The same fish, larger but just as beautiful, were at Bloody Falls on the Coppermine, but the force of the big river was too much for any fish, and they rose quietly in the eddies. I thought of Samuel Hearne, certain that the mighty rush of water through the rock gorge remained in his mind's eye to the end of his days as it will in mine, though for different reasons.

Lower Islands pool, Campbell River, British Columbia

Roderick Haig-Brown fishing the Campbell River, British Columbia, September 1976

I have caught speckled trout in many places, from British Columbia and Idaho to Argentina, but never in eastern Canada, his true home. But the search for big lake trout in shallow bays and along the edge of the ice in spring breakup, with the Canada bird singing from the tips of the little spruce trees, is not easily forgotten; and it is an essential Canadian experience to catch, just once, a muskie from the bright waters of Georgian Bay.

I have fished through the ice, alone and not very effectively, once or twice. But I can imagine the warmth and good fellowship of the fishing huts on Lake Simcoe or Lac St. Pierre.

It is easy to dream of fishing in Quebec and the Maritimes. A week on the Jupiter River on Anticosti Island perhaps, a month testing rivers along the west coast of Newfoundland or some pleasant summer days on Cape Breton Island, inquiring here and there about the chance of a salmon or some speckled trout. After all, salmon come to a floating fly more readily in Canada's Atlantic provinces than anywhere else on earth. It would be a happy experience to search for a five-pound sea-run speckled trout in the rich estuaries of Prince Edward Island and think perhaps of trying next day for a six-hundred-pound bluefin. Yet most anglers do not need or want all these experiences. It is enough and more than enough to be able to return again and again to the familiar places and find the fish still there.

The philosophy of it all? Perhaps it can be summed up rather simply: The resource is a trust and the first responsibility of angler, manager, scientist and politician is to ensure its protection and perpetuation. Others who come after us will need it. Following upon this, a first concept is that the resource belongs to the people of Canada, but the sport, its values, traditions, standards and ethics, belongs to the anglers themselves and is in the care and keeping of anglers everywhere. The value of it all? It is worth as much or as little as people find in it: as little as an hour or two of happiness in a small child's day (if that is little) or as much as a long lifetime of happiness and sophisticated contentment. Measured in terms of the millions of men, women and children who turn to angling for the pleasure that comes from active participation in the world's true wealth of unspoiled natural things, this is happiness in massive amounts, harming no one, benefitting everyone. And the pleasure and happiness of bright fish in bright waters has deep meaning for many people who are not anglers at all.

21 September 1976
Campbell River, B.C.

Bibliography

Adamson, W. A. *Salmon Fishing in Canada.* Edited by Col. Sir James Edward Alexander. London: Longman, Green, Longman and Roberts, 1860.

Berners, Dame Juliana. *Treatise of Fishing with an Angle.* 1496.

Blake, W. H. *Brown Waters.* Montreal: The Reprint Society of Canada, 1948.

———. *In a Fishing Country.* Toronto: Macmillan Company of Canada Ltd., 1922.

Clemens, W. A., and Wilby, G. V. *Fishes of the Pacific Coast of Canada.* 2nd ed. Fisheries Research Board of Canada, Bulletin no. 68. Ottawa: Queen's Printer, 1961.

Comeau, Napoléon A. *Life and Sport on the North Shore.* 3rd ed. Quebec Telegraph Printing Company, 1954.

Halford, Frederic M. *Dry-fly Fishing in Theory and Practice.* 2nd ed. London: Sampson Low, Marston, Searle and Rivington, 1889.

Herbert, Henry William. *Frank Forester's Fish and Fishing of the United States and British Provinces of North America.* London: R. Bentley, 1849.

Jordan, David Starr, and Evermann, Barton W. *American Food and Game Fishes.* New York: Doubleday, Page and Co., 1902.

La Branche, George M. L. *The Dry Fly and Fast Water.* New York: Charles Scribner's Sons, 1914.

Lambert, T. W. *Fishing in British Columbia.* London: H. Cox, 1907.

Miller, Richard. *A Cool, Curving World.* Toronto: Longman Canada Ltd., 1962.

Norris, Thaddeus. *American Angler's Book.* Philadelphia: Porter & Coates, 1864.

Palmer, C. H. *Salmon Rivers of Newfoundland.* Boston: Farrington Printing Co., 1928.

Radcliffe, William. *Fishing from the Earliest Times.* London: John Murray, 1921.

Scott, W. B., and Crossman, E. J. *The Freshwater Fishes of Canada.* Fisheries Research Board of Canada, Bulletin no. 184. Ottawa: Queen's Printer, 1973.

Skues, G. E. M. *Minor Tactics of the Chalk Stream.* London: A & C Black Ltd., 1910.

———. *The Way of a Trout with a Fly.* London: A & C Black Ltd., 1921.

Walton Izaak. *The Compleat Angler.* Rich, Marriot, 1653.

Weeks, Edward. *The Moisie Salmon Club, A Chronicle.* Barre, Mass.: Barre Publishers, 1971.

Illustration Credits

Atlantic Salmon Journal
p. 26 *(bottom)*
Den Austin
p. 87
Canadian Government Office of Tourism
pp. 20, 24–25, 30 *(top and bottom)*, 35, 40, 42 *(bottom)*, 45, 46 *(top)*, 51, 52, 56–57, 58, 67, 68 *(top and bottom)*, 71, 72–73, 78, 84, 88, 93, 99 *(top and bottom)*, 104–105, 110, 115, 120 *(bottom)*, 122, 125, 126, 131 *(top)*
Department of Fisheries and Oceans, Canada
p. 94
Robert Fish
chapter titles, pp. 1, 19, 36, 39, 77, 100, 103, 116, 119 (p. 19 *bottom* and p. 100 *top* courtesy of Dr. H.G. Wadman)
George Gruenefeld
pp. 26 *(top)*, 29, 90 *(top and bottom)*, 132
Roderick Haig-Brown
pp. 136-137
Richard Olmstead
pp. 14, 46 *(bottom)*, 62, 83, 109, 131 *(bottom)*
Christopher Springmann
front and back of jacket, p. 138
Travel Manitoba
pp. 42, 61, 120 *(top)*

Fly drawings by Robert Fish

Thunder and Lightning pp. 15,127
Atlantic salmon fly pp. 18,95
Durham Ranger p. 31
Green Highlander pp. 41,70
Hair Wing salmon fly p. 50
March Brown p. 63
Dusty Miller p. 80
Lady Caroline p. 111